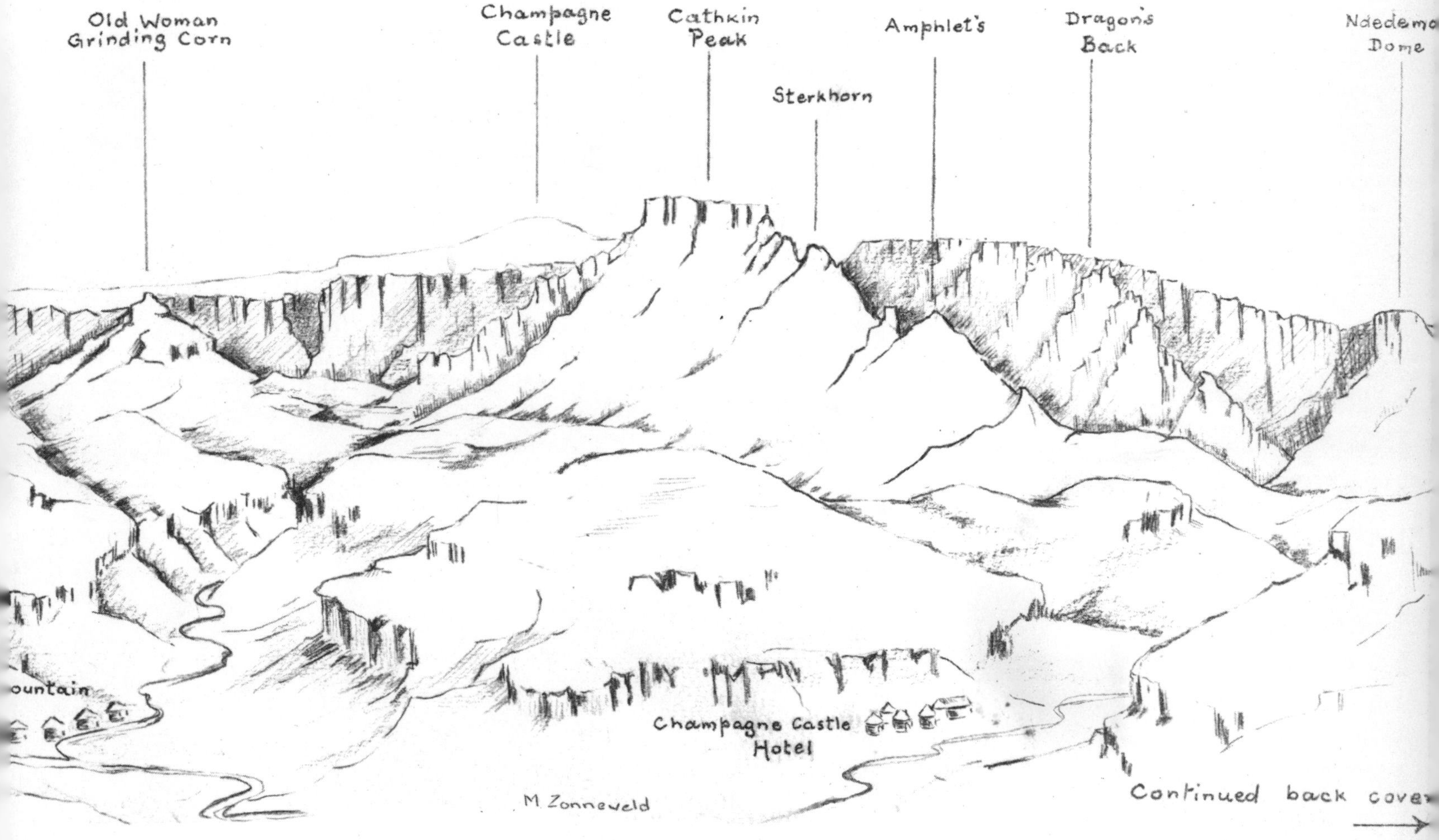
Old Woman Grinding Corn
Champagne Castle
Cathkin Peak
Sterkhorn
Amphlet's
Dragon's Back
Ndedema Dome
ountain
Champagne Castle Hotel
M. Zonneveld
Continued back cover

MOUNTAIN SPLENDOUR
WILD FLOWERS OF THE DRAKENSBERG

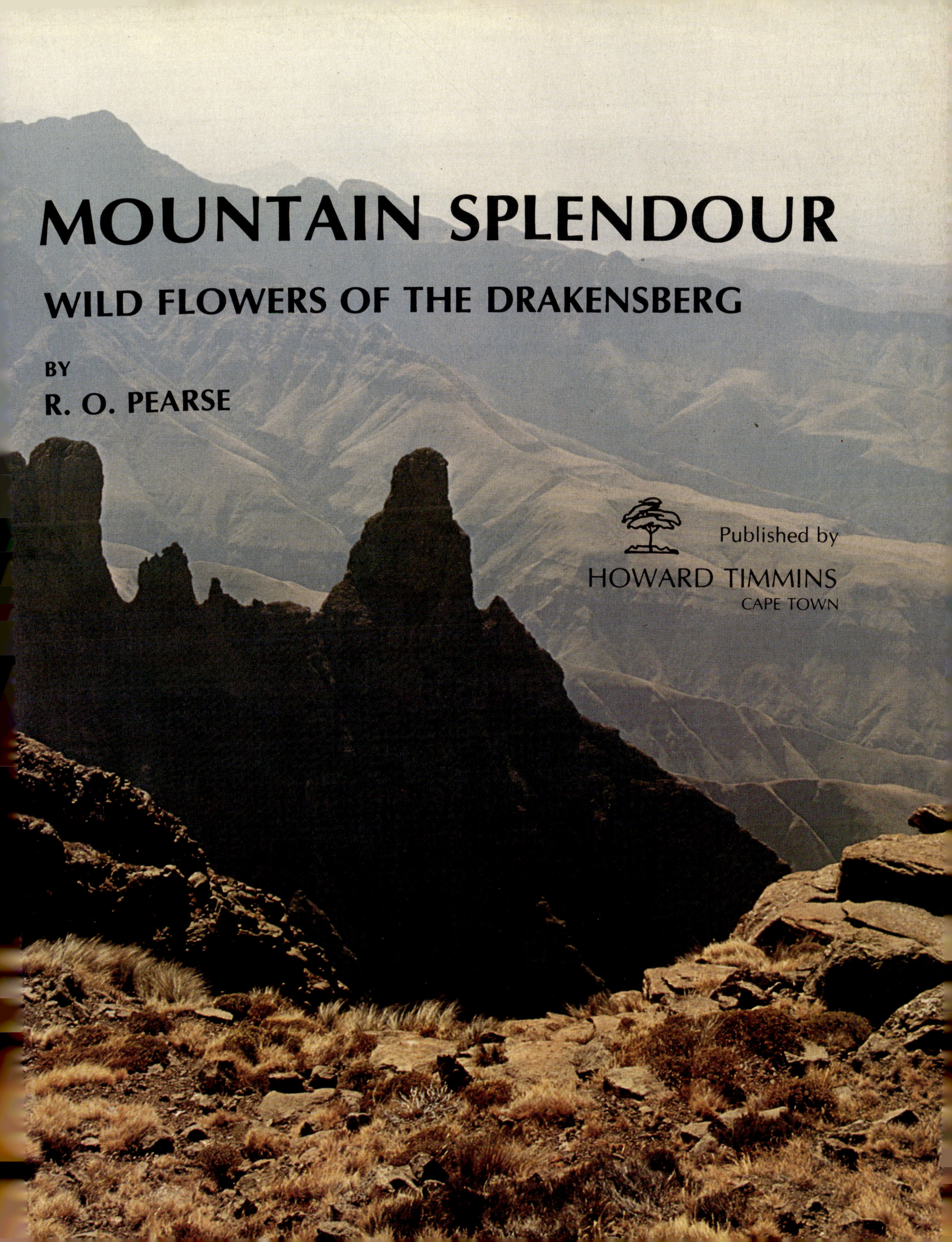

MOUNTAIN SPLENDOUR

WILD FLOWERS OF THE DRAKENSBERG

BY

R. O. PEARSE

Published by

HOWARD TIMMINS

CAPE TOWN

Howard Timmins (Pty) Ltd
45 Shortmarket Street
Cape Town

First Edition 1978

Photosetting, Layout and Reproduction by McManus Bros. (Pty) Ltd. Cape Town.

Lithographed and bound by ABC Press (Pty) Ltd Cape Town.

ISBN 0 86978 156 1

Previous page: Mountain Splendour. In the foreground, Column and Pyramid. On the skyline, the Cathedral Range.

CONTENTS

Owing to the high costs of colour reproduction it would have been impossible to publish this book at a reasonable price without the generous help of a number of people. We wish to express our gratitude to the following individuals and organizations, all of whom have offered either generous financial assistance, or help in other ways:

Toyota South Africa Limited

Mr. A. W. van der Riet, of Cathedral Peak Hotel

The Natal Building Society

The Drakensberg Boys Choir School

The Wildlife Society of South Africa, Estcourt Branch

Mr. John Tungay

*For Edith
who loves all
lovely things*

TRIBUTE

The mountain flowers of the Drakensberg – many seldom seen by human eyes – are now presented to the world in this superb book by Reg Pearse, famous educationalist, author and photographer.

This meticulous task took him nearly 30 years of observation, research and photography over a wide area and to the very pinnacles of the towering peaks, and it covers all the seasons of the year.

"Mountain Splendour", his second contribution to the cultural heritage of our beautiful country, is an excellent book which I can heartily commend.

P. G. J. Koornhof

Minister of Education, Sport and Recreation

Below: Sentinel Peak under a heavy fall of snow.

FOREWORD

In this book Mr R. O. Pearse passes on to us more of his unrivalled knowledge of the Natal Drakensberg. Study of this region has been his favourite form of relaxation for over half a century. He has walked over the whole area in all seasons, climbed most of its highest peaks, studied its physical features and scenery, and has become familiar with the flora, fauna and peoples. In the present book we can share with him his love of the high altitude grassland flowers.

Those who are acquainted with his earlier book, "Barrier of Spears", or who have listened to his inspiring illustrated lectures, will realise that both he and his son Malcolm are artistically skilled photographers, and will look forward to seeing how successful they have been in the rather specialised field of plant photography. It is, of course, impracticable in a single book of this kind to illustrate all the flowers, but Reg Pearse has endeavoured to provide a selection of species from the most striking genera. In addition to the illustrations he gives the main habitats, the differences between the species, some common names and what he has discovered about the medicinal and other uses.

With reference to the medicinal uses the Introduction, very wisely, contains a warning that many of the plants used as Zulu medicines are highly toxic. It must be emphasised that the lore of crude tribal medicines has no strict scientific basis; most are harmful or of no therapeutic value. There is as little reason to resort to their use as there is to revert to the herbal treatments and quackery of the Dark Ages, even though the information given is of considerable sociological interest.

Common names of plants, whether in English, Afrikaans or Zulu, are also not based on any scientific principles. Many arise fortuitously, are often of purely local use, and are sometimes applied haphazardly. Naturally they have their value, being easily remembered, easily understood, not requiring knowledge of any strange language, and occasionally very apt. Their origin and derivation may also be etymologically interesting and intriguing. The number of such names in a language gives some idea of the attractiveness of a flora and of the extent of the interest which people take in it. It is, for instance, quite remarkable how numerous are the Zulu plant names, due, no doubt, to the Zulus having lived for some centuries close to nature in a country with a luxuriant sub-tropical vegetation.

Reg Pearse's services in the educational and spiritual fields which were the main concern of his long and valuable professional career are well-known and highly respected in Natal. It is a pleasure for me to pay tribute to them and to express appreciation of the present and other contributions he has made to interest others in the splendours of the Drakensberg.

A. W. Bayer
Professor Emeritus of Botany, University of Natal

INTRODUCTION

This book has been written for the average lover of wild flowers, and particularly for the hiker and mountaineer who desires nothing better than a walk in the sunshine, with the towering peaks around him and the wild flowers at his feet.

I am not a professional botanist, and I have had little formal training in botanical matters. But for nearly 40 years I have climbed the high peaks of the Drakensberg, usually in the company of my son Malcolm, and over the years we have grown to love the flowers of this magnificent mountain region, and always we have carried our cameras with us. Between us we have accumulated over 1 000 transparencies of some of the loveliest flowers in all South Africa. This book is the result of these efforts, and it is offered to the public in the hope that it will help to awaken an interest in our mountain flora, and a desire to ensure its preservation. The French have a motto "to know all is to forgive all". If that is so, to know better is also to love better and I believe that the more we know about these mountain splendours the more we will grow to love them, to cherish them and to preserve them.

One word of warning. The Cape is famous for its wild flowers. There you will find whole fields of the most exquisite blooms, in massed array. Here, in the Drakensberg, you rarely find those great splashes of colour that burst upon your eyes in the Peninsula, or on the plains of Namaqualand. You will often have to hunt diligently for the Drakensberg wild flowers, and sometimes they are small, and you might pass unheeding by. But your search will be well rewarded. And may I remind you of the words of one who loved the daffodils of Ullswater:

To me the meanest flower that blows can give
Thoughts that do often lie too deep for tears.

Photographing flowers in the Drakensberg is not easy. It demands unlimited time and plenty of patience. The terrain is steep. Places where the flowers grow are often difficult to reach, and many species are tucked away on almost inaccessible mountain ledges. It has often taken me a full day to climb to the spot where a particular species was growing, carrying, in addition to heavy photographic gear, sleeping bags in which to spend the night before descending the next day. Added to all this are the tearing winds that sweep the high places, making time exposures impossible, and the constant mist and drizzle that one gets in summer, the very time of the year when so many of these mountain plants are flowering at their best. On several occasions I embarked on a three-day trip into the mountains to photograph a particular species, only to find on my arrival at the locality that I was too late. The flowering period was over, and I had to wait a full twelve months before my chance came round again. In the final analysis it has taken me something like 25 years to obtain the pictures in this book.

It would be quite impossible to include every genus and every species in a book of this nature. In the early 1950's Dr D. J. B. Killick, Deputy Director of the Botanical Research Institute, Pretoria, made a detailed study of the ecology of just one small area in the Drakensberg, the Cathedral Peak area, and he listed 419 genera and 907 species of flowering plants. In the whole Drakensberg range the total must be much greater than this. In making my selection I have been guided by two principles. I have tried to make my selection as representative a cross-section of the flora of the area as possible, and I have also tried to include the more spectacular flowers, and those most likely to attract the notice of visitors. All the flowers in this book are indigenous except for a few exotics which I have felt it wise to include, as these are often noticed by climbers and ramblers. These plants, introduced many years ago from overseas by the early settlers, or sometimes growing from seed included in imports of grain from other countries, have now established themselves in our flora and have spread to the countryside. In all such cases I have made a note of the fact that they are exotics.

The book has been written in a language which I hope the layman will understand, and I have cut down as far as possible the use of erudite botanical terms, but the use of some was inevitable. The whole book has been checked by experts for scientific accuracy, and to assist the reader a Glossary of botanical terms has been included on page 233.

I am particularly grateful to Professor A. W. Bayer, Emeritus Professor of Botany at the University of Natal, who has read the whole MS through, who has made a number of very necessary corrections, has offered innumerable most useful suggestions, and who has so graciously consented to write the Foreword.

Plant families have been arranged according to the system of de Dalla Torre and Harms. This is the system in use in the National Herbarium, Pretoria; in the Natal Herbarium, Durban; and in the University of Natal

Herbarium, Pietermaritzburg. Individual genera and species, however, have not been arranged in any particular order, the exigencies of lay-out being the main consideration.

Topography and Climate The Drakensberg forms part of the eastern escarpment of Southern Africa, and extends from the Eastern Cape to the Northern Transvaal, a distance of roughly 970 km. This book deals only with the Natal Drakensberg, where it forms the boundary between Natal and Lesotho, a distance of about 180 km.

The geology of the Drakensberg fortunately is fairly straightforward. It consists of two clearly-defined areas, firstly the summit area and the gigantic wall of the main Drakensberg, rising from about 2 000 metres to 3 000 metres and more. This is a basalt area, of volcanic origin. These towering peaks and steep mountain slopes are superimposed on the second area, mainly of sandstone. This is an area of somewhat gentler slopes, stretching from the floor of the river valleys at about 1 250 m to about 2 000 m, capped with several layers of cave sandstone and known as the Little Berg. These two 'layers', the Sandstone and the Basalt, are consistent throughout the length of the Natal Drakensberg. (For a more detailed account of the geological origins of the Drakensberg, see the Prelude to the writer's book, "Barrier of Spears", published by Howard Timmins of Cape Town.)

The main Drakensberg, also known as the High Berg, is a bleak and somewhat harsh area, often snow-covered in winter, but with an amazing wealth of floral beauty, often hard to believe. The Little Berg is more benign and gentle, with richly-grassed mountain slopes. The forests are deep and cool, full of the sound of insects, topped at the 1 800-2 000 metre level with a fantastic series of tremendous sandstone cliffs.

It is a region of secret valleys and of mountain streams, tumbling down from the heights in the clearest water to be found anywhere in South Africa.

These two areas result in three main terraces in the Drakensberg, the floor of the river valleys in the Little Berg at about 1 250 metres, the summit of the Little Berg at about 2 000 metres, and the summit plateau of the main Drakensberg at about 3 000 metres. Dr D. J. B. Killick, in his "Account of the Plant Ecology of the Cathedral Peak Area" (1963) recognises three distinct vegetation belts in the Drakensberg, corresponding roughly to these three terraces.

The first of these is the Montane Belt, extending from

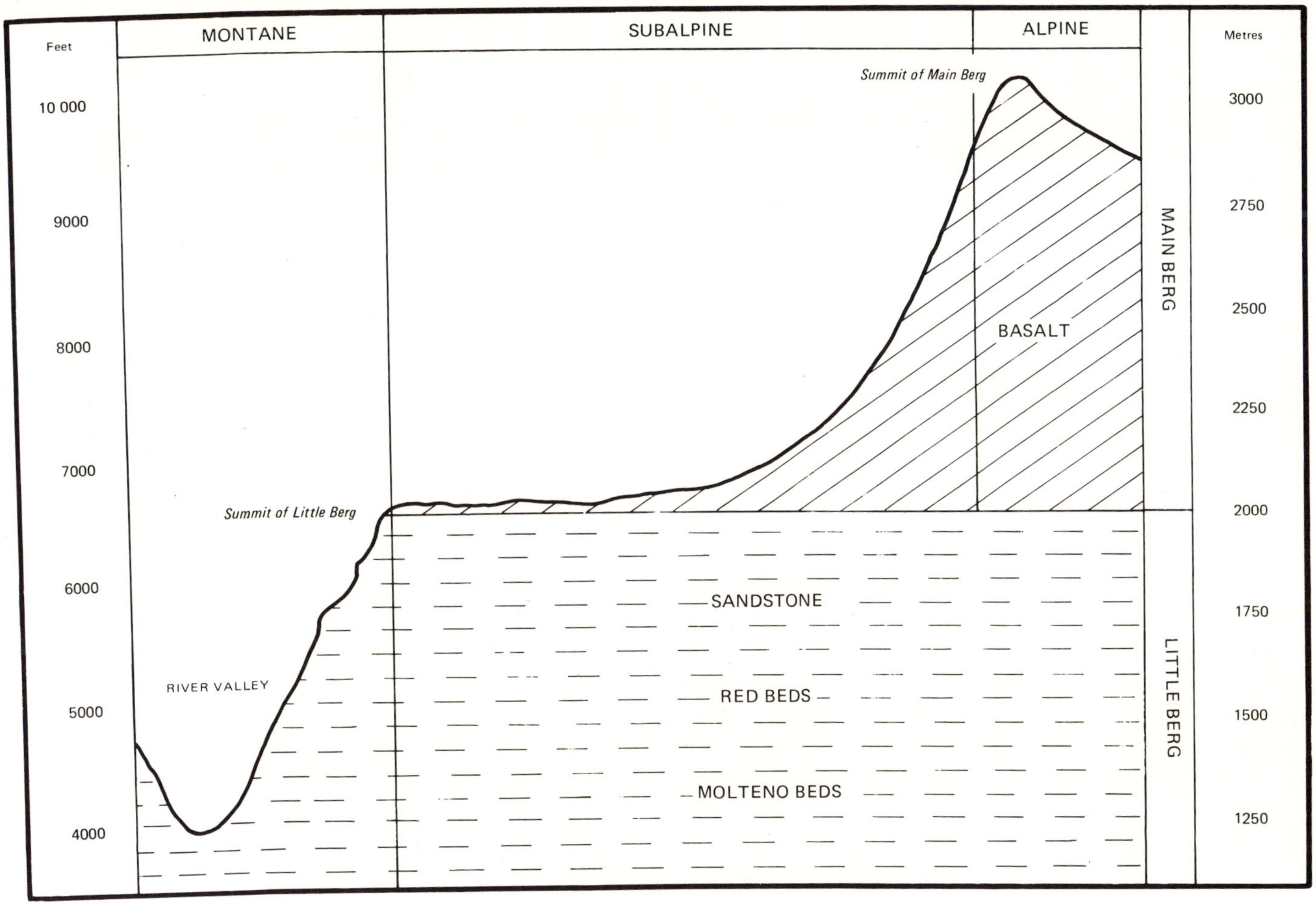

PROFILE THROUGH DRAKENSBERG

showing main topographical details and the three vegetation belts.

the valley floors at 1 250 metres to the summit of the Little Berg at 2 000 metres. The area consists mainly of the slopes of the Little Berg and the river valleys. Rainfall is about 1 400 mm per annum. The greater part of this area is occupied by grasslands, chiefly *Themeda triandra,* with species of *Protea* scattered on the hillsides. The upper level, 2 000 metres, is also the limit of the tree-line, and many of the river valleys are thickly forested, mainly with yellowwoods, *Podocarpus latifolius, P. falcatus* and *P. henkelii.*

The Subalpine Belt extends from the summit of the Little Berg to just below the summit of the Main Berg. Rainfall is high, an average of 2 000 mm per annum, and the mountain slopes steep. The area lies above the forest line, the only woody plants being *Leucosidea sericea,* the umtshitshi bush. As in the Montane Belt, the vegetation consists mainly of grasslands, with *Themeda triandra* the dominant grass, but also containing temperate grasslands as well.

The Alpine Belt occupies a narrow strip at the edge of the Drakensberg escarpment, extending down into Natal to about 2 865 metres and into Lesotho to about the same altitude. (From the edge of the escarpment the ground slopes down gradually westward, into Lesotho). Rainfall is less than in the Subalpine Belt, about 1 600 mm per annum. Here the climate is severe, and long droughts are not unknown. The ground is tundra-like, often boggy and swampy in summer, with shallow soils. There is a decrease in atmospheric pressure and an increase in the intensity of ultra-violet rays, resulting in a decrease of the growth-promoting hormones and a consequent dwarfing of many of the plants growing at these high altitudes. Even so, the profusion of flowering plants on the summit, especially in summer, is remarkable. (Hans Justus Thode, pioneer plant collector in the Drakensberg in the 1890's, described the vegetation on the summit as "a veritable flower-garden"). Another noteworthy feature of these high altitudes is the extreme range of temperatures, from very cold at night to very hot at midday. Humidity too varies greatly being high in summer, due to the abundance of cloud and mist, and low in winter. Snow is frequent, mainly in winter, but I have known snow to fall, rarely, at some time or another, in every month of the year. In winter it often lies for months at a time on the summit plateau and in south-facing gullies. Interesting, though, is the fact that this blanket of snow protects plants from excessively low temperatures, and prevents the soil from freezing.

Elsewhere in the Drakensberg, especially in the valleys of the Little Berg, the climate is mild and pleasing, cool in summer and certainly warmer in winter than in the Midlands of Natal. Summer is the rainy period, with mist and rain sometimes lasting for several days. Autumn and winter are brisk and bracing, with day after day of cloudless sunshine. Spring can sometimes be unpleasant, especially when the rains are late and the hot "Berg" winds blow from the northwest.

In the text of this book I have tried not to confuse the reader by using too many technical terms. I have therefore called the Montane Belt the Little Berg, and the Subalpine and Alpine Belts I have grouped together under the more commonly accepted terms High Berg or Main Berg.

Fires. Plant life in the Drakensberg is affected not only by altitude, vegetation regions and climate, but also by fire. Grass fires are common in the Drakensberg, especially in the winter and early spring months, and they have a profound effect on the vegetation.

It must be remembered, first of all, that although grass fires are much more common today than in the early days, they are not entirely man-made, and have undoubtedly been known in the Drakensberg for thousands, if not millions, of years. Fires are often started by lightning and by sparks from rocks rolling down the mountain side. The vegetation has had to learn to adapt itself to these periodic fires.

At the same time it must be admitted that with the advent of man fires are much more frequent to-day than they were in the olden days. Undoubtedly some of our mountain species are threatened by these raging fires which sweep down, year after year, from the heights.

Seventeen years ago, just prior to our retirement, my wife and I bought a 6 ha plot in the foothills of the Drakensberg, on a sloping piece of ground on which there was not a single Erica to be seen. We fenced it, and for seventeen years the hillside above our home has not been burnt at all. To-day that hillside is covered, every March and April, with masses of *Erica woodii* flowers, a great sheet of magenta-pink, while just over our boundaries, where the grass is burnt every year, not a single *Erica woodii* plant can be found. It would seem that the plant literally disappears with frequent burning.

Erica cerinthoides provides another interesting example. It is found sparsely throughout the Drakensberg, but the plants never reach a height of more than 0,25 metres. In the Wild Flower Garden in the Giant's Castle Game Reserve, however, where the plants are protected from fires, there are two bushes of this Erica which flower profusely every year, and which have reached a height of over a metre. But you will notice one interesting phenomenon. Those plants in the wild,

The slopes of the Little Berg, showing the sandstone rock-bands, the mainly Themeda triandra *grasslands, dotted with* Protea *species, and above, the steep basalt slopes of the Main Drakensberg (Cathkin Peak and Sterkhorn).*

which are subject to periodic burning, bear larger flowers, and of a brighter red, than those which are never burnt.

Proteas, too, are adversely affected by fires. *Protea roupelliae* grows in grasslands on the slopes of the Little Berg, but it also grows in rocky outcrops and on the cave sandstone plateaux. Its thick bark protects it to some extent from the damage caused by fires, but you will find that those trees growing in rocky outcrops, where the grass cover is thin, are far finer specimens than those growing in grasslands, where the heat from the fire is so much more intense. I recently came across a magnificent stand of *Protea roupelliae* at the head of Bezuidenhout Pass, growing on a massive outcrop of cave sandstone, where few fires could ever reach them.

W. R. Trauseld has told me that he has noticed that *Protea multibracteata,* with its thick, corky bark, is able to withstand the effect of fires, but if the fire comes late in the season, when the sap is flowing freely, the trees often have burst stems, due, he thinks, to the fact that the sap vaporises and, being confined, bursts the stem.

This is not the full story, however. Occasional fires do act as a regenerating factor, and I have noticed that in those areas of the Drakensberg which are subject to *occasional* fires, the flush of early spring flowers is far finer than on our own unburnt plot, where the grasses have grown coarser and the veld flowers are smothered. The brave show of new flowers in the blackened veld of early spring, after a fire, is a well-known sight in the Drakensberg.

Unquestionably, too, the germination of some seeds is stimulated by fire. The silver wattle, *Acacia dealbata,* is a well-known example of this. *Protea subvestita* is another. The two other Protea tree-species, *Protea roupelliae* and *Protea multibracteata,* protect themselves against fire by means of their thick bark. *Protea subvestita* lacks this, and the trees protect themselves partly by growing much more thickly together than in the case of the other two species. But the tree also ensures its survival in much the same way that the wattle does. You will often see a lone *Protea subvestita* blackened and killed by the fire, but around it are literally hundreds of young plants growing from seed stimulated by the very fire which killed the parent plant. Mr W. R. Trauseld believes that in the case of *Protea subvestita* the hairs on the seed get burnt, thus allowing the seed to come into closer contact with the ground, permitting better germination, whereas with the wattle, as he rightly points out, the heat of the fire causes the hard epidermis of the seed to crack, in this way allowing moisture to reach the embryo and so start germination.

In our Drakensberg areas plants have adopted many methods which enable them to survive fires. That is why you will so often find Ericas growing in south-facing rock gullies, where the fires rage less fiercely than on north-facing slopes. Bulbous plants are more easily able to survive a fire than plants equipped with roots, especially shallow roots. This is undoubtedly one reason why the Drakensberg is so rich in flowers from the Liliaceae, Iridaceae and Amaryllidaceae families, most of which have bulbs or corms.

We may conclude, then, that many plant species in the Drakensberg are gravely endangered by mountain fires, that many have adapted themselves to withstand the effects of fires, but that *occasional* fires are beneficial, acting as a regenerating factor, clearing out the thickly-matted old grasses and dead undergrowth, and stimulating seed germination.

Nomenclature. Flowers are given scientific names, but they also have common names. These latter are often interesting, but as a general rule they should not be used. They are ambiguous, misleading and often cause confusion. One reason for this is that they often vary from district to district, and from country to country. Sometimes a common name is applied to more than one individual flower. I know, for instance, of at least

Below: Mountain streams tumble down from the high peaks through patches of forest into the valleys of the Little Berg. On the left is a Greyia sutherlandii *in bloom, and next to it,* Leucosidea sericea.

three different flowers called "Black-eyed Susan" – *Thunbergia alata, Hibiscus trionum* and *Callilepis laureola*. Conversely, a single flower can have more than one common name. *Haemanthus magnificus* has at least six common names, Blood Flower, Torch Lily, Paint Brush, Snake Lily, Varkoor and April Fool. In any case South Africa is too young a country, only about 300 years old, to have built up, as in England, a common country lore in this respect.

Originally plants were identified by botanists by a series of long and complicated Latin words which, while often descriptive, made no attempt at indicating the affinities between different plants. But in 1737 the Swedish botanist Linnaeus (1707-1778) simplified this. He persuaded the botanical world to adopt the system whereby plants should have only two names, first a generic name ('genus', plural 'genera') and secondly a specific name ('species'). This is known as the binomial system, and it has been accepted worldwide. The generic name is like your surname. It indicates a group of plants all with certain common characteristics, e.g. *Protea, Gladiolus, Watsonia*. The specific name is like your Christian name e.g. Watsonia *socium*. Each genus contains a number of these different species, plants which differ only very slightly from one another, and which are capable of cross-breeding with each other. (A species of one genus rarely cross-breeds with a species of another genus. The gap between them is too large). Thus we have *Watsonia densiflora* and *Watsonia meriana,* and so on. Some genera have hundreds of different species. Once the binomial name has been given the plant is known by that name throughout the world.

The different genera, in turn, are now grouped into families, so that we have, for instance, the 'Daisy' family, the Compositae; the 'Bean' family, the Leguminosae, and so on, each with sometimes hundreds of different genera and possibly thousands of species.

Most botanical names are either Greek or Latin in origin, though some are derived from people and places. Almost always they are descriptive of some aspect of the plant. Sometimes people are rather afraid of these Greek and Latin names and are resistant to them. The names appear long and complicated. But they are the only *safe* way of naming plants, and every effort should be made to become familiar with them. And this is not difficult. In any case, we use many of these names almost every day of our lives. Gladiolus, Watsonia, Nemesia, Protea, Nerine, Disa, Chrysanthemum, Petunia, Dahlia and Pelargonium are all scientific names, and there are many more. There is nothing intrinsically difficult, mysterious or strange about such names.

Unfortunately, as our knowledge increases, it is often found necessary to change the name of a plant. Sometimes the plant is found to have been placed in the wrong genus. Sometimes it is discovered that a particular plant had been named differently years before by someone else, in which case the first name becomes the correct name. These changes, especially when they apply to well-known plants, old friends whom we have known for many years, can be most irritating, but in our irritation we should remember that they are very necessary, if our nomenclature is to be kept scientific, neat and tidy. The rules of nomenclature have now been internationally accepted, and are used throughout the world.

They can be very complicated, but basically there are only three that need concern us:

1. The earliest name given to a plant after 1753, the year in which Linnaeus published his "Species Plantarum" is the one that must stand. If, for instance, a plant was given a certain name in 1920, and it is subsequently discovered that someone else had already named the plant a hundred years earlier, then the 1920 name must fall away.
2. If the name of the genus is changed, the earliest *specific* name must be retained. Flowers are like women. Their surnames can change (if they marry), but they always retain their Christian ('specific') names. (There is an exception here. In a few rare instances it was found that when a plant was placed in a different genus and kept its specific name, this specific name had already been used for another plant in the genus. This would have meant two different plants with the same name, and a new specific name had then to be found for the new entrant into the genus).
3. Once a name has been given to a plant it can never be used for another plant, even if this name has fallen away and been discarded through the application of Rule 1 above.

Scientifically the two binomial names of a plant should be followed by a third 'name' which is always given as a capital letter or an abbreviation. This is the name of the botanist who described and named the plant, the 'author' of the plant. Throughout this book I have omitted this third term, as it is unnecessary for the general reader.

Zulu Medicinal Usages. From the very earliest times there has been a close connection between plants and the practice of medicine. So close was this relationship that in the Middle Ages and early Renaissance periods the Professor of Medicine at a University was almost always the Professor of Botany as well. Even to-day many of our proven remedies come from plants. Quinine is but one example of this. In Africa the native races, and in particular the Zulus, the Xhosas and the Sothos have not been found wanting in this respect, and there are few of our wild flowers that do not have interesting uses amongst Africans. The curative properties of some plants are quite genuine. Others are of doubtful value. Many plants are connected with witchcraft and superstition, but all these usages are interesting, and in this book I have tried to preserve as many as possible.

Several words of caution, however, are necessary. It must be remembered that superstitious usages apply almost *in toto* to the unlettered kraal Zulu or Xhosa. These usages are rapidly dying out amongst the more educated and sophisticated Africans.

It must also be remembered that the medicinal uses of a plant can vary enormously from tribe to tribe and from place to place. One has to be particularly cautious in making generalizations in this respect, and some usages can be mere hearsay. In all these matters I have tried to be guided by the monumental work of Watt and Breyer-Brandwyk, "Medicinal and Poisonous Plants of Southern and Eastern Africa".

But the most important *caveat* concerns the actual usage of alleged 'medicinal' plants. It must be remembered that the average Zulu herbalist is usually illiterate and has had no training in botany, pharmacology or therapeutics. On no account should the layman experiment in any way whatever with the supposed curative properties of South African plants. Many are highly toxic, even in small doses. Others, which have proved therapeutic values in small doses, can quickly kill if the dose is exceeded. Professor A. W. Bayer tells of a case in which he was personally involved. This was of a New Hanover Zulu mother. She had three children, and had recently lost her husband. She wished to protect them from the evil spirits which, having killed her husband, might have designs upon the children. She gave them a meal of the tips of a wild cucurbit. All three children died. Recently the South African Medical Journal has had several trenchant articles on the subject. The reader is referred to page 204 of this book, where the deadly properties of *Callilepis laureola* are discussed, properties which have recently led to a series of deaths at King Edward VIII Hospital in Durban. Another example is the small shrub with delicate, lace-like mauve flowers known to Zulus as iboza *(Iboza riparia).* Here the leaves are used by the Zulus as a remedy for colds and influenza, but if the traditional dose is exceeded death rapidly ensues. In patients who died after using Iboza, urine and stools consisted of almost pure blood, mixed with shreds of skin. Many other examples of traditional Zulu remedies leading to fatalities could be quoted. (See also page 20, where fatalities occurring from the ingestion of *Encephalartos* seeds are referred to).

African Names. Wherever possible I have given the Zulu, Xhosa or Sotho names of plants, as far as they are known to me. But here again caution is required. Some names apply only to small areas, and are not universally accepted. Here I am most grateful to Mr S. Bourquin, the Director of the Port Natal Bantu Affairs Administration Board, who has placed his unrivalled knowledge of Zulu orthography, customs and folk-lore at my disposal. He has checked the spelling of all the Zulu names, and has made a number of most useful comments.

Wild Flower Protection Act. It is essential for the general public to realise what a priceless heritage we have in our wild flowers, and for all of us to do what we can to protect them. Apart from their glorious beauty, they are a unique, and a diminishing, asset – unique if only that, apart from a few parasites and semi-parasites, flowers are the only living things that can exist without drawing nourishment from other living things, and diminishing, because of the encroachment on their environment of building developments and the ever-widening network of roads and railways. An awakening interest in our wild flowers (and the promotion of this is one of the main reasons for this book) is all to the good, but this awakening interest should not be an invitation to go into the mountains and to strip them of their treasures. In any event, this is illegal, and you would render yourself liable to a heavy fine if you were caught. Rather than this, photograph them in their natural habitat, and leave the plants themselves untouched. Better still, try growing some of our wild flowers in your garden. In this way you are actually helping to preserve them. But when you do this, do refrain from digging the plants up to obtain specimens. Remember again, this is illegal. It is also illegal to collect seed in the open. You should rather obtain seed from Kirstenbosch (members of the Botanical Society are entitled to free issues of seed) or from one of the many authorized nurseries which are permitted to do just this.

All four Provinces of the Republic have their own Wild Flower Protection Ordinances, and it would be wise to obtain a copy of the Ordinance in respect of your own Province if you want to sell indigenous plants or if you wish to gather them for any purpose

The summit of the Little Berg, a region of flat, rolling grasslands, rocky outcrops, and with the peaks of the Main Drakensberg towering above. In the foreground is a plant of Rhus dentata, *and next to it is a Cycad,* Encephalartos ghellinckii.

whatever. Penalties for non-compliance with your particular Ordinance can by heavy.

In Natal, the Ordinance relating to Wild Flowers is Chapter XI of the Nature Conservation Ordinance No. 15 of 1974. (The Ordinance may be obtained from Lex-Patria Publishers, P.O. Box 8161, Johannesburg, 2000). The Ordinance classifies indigenous plants into three categories:

A. Unprotected plants, such as grasses, bracken etc. These are not subject to any control.

B. Protected plants, which includes all indigenous species except those included in A and C.

C. Specially protected plants, which includes those species specially in need of extra protection by virtue of their rarity or because there is a demand for them for commercial purposes. The various species falling into this category are listed by proclamation from time to time.

Control in the case of B and C above includes the gathering of the various species, their selling, their purchase and their export.

Two provisions which we should particularly note are:

1. No indigenous plants may be purchased from or sold to anyone without a permit.
2. No indigenous plants may be gathered anywhere, including roadsides, without a permit, unless the land is your private property.

In Natal permits for the gathering of wild flowers are granted to bona fide collectors who require the plants for scientific purposes, by the Natal Parks Board (address P.O. Box 662, Pietermaritzburg, 3200).

Penalties for non-compliance with the Ordinance can be as heavy as a fine not exceeding R500 or imprisonment not exceeding twelve months, or both. They can also include forfeiture of the vehicle used in committing the offence.

But surely there is a better way than being motivated by fear of the law. Let us rather know our wild flowers, and thereby grow to love them, for if we love them we will also cherish them and desire their protection. I appeal to all readers of this book to do just that. When Francis Bacon wrote his *Essay on Gardening* in 1597, his opening words were: "God Almighty first planted a garden". Man lost his Eden through his own folly. God has also planted a garden of exquisite beauty in the Natal Drakensberg. Are we to lose this as well?

"Emkhizweni",
Private Bag,
Winterton.
April, 1978.

R.O.P.

These I have loved:

Radiant raindrops couching in cool flowers,
And flowers themselves, that sway through sunny hours,
Dreaming of moths that drink them under the moon.

Rupert Brooke

WILD FLOWERS OF THE DRAKENSBERG

ENCEPHALARTOS *Zamiaceae*

The family of the Zamiaceae includes the plants commonly known as Cycads. Strictly speaking we should not have included Cycads in a book on wild flowers, for they are not true flowering plants: their seeds are not enclosed in an ovary, but form a cone in the centre of the plant. Visitors to the Drakensberg, however, are fascinated by these strange plants, and for that reason *Encephalartos ghellinckii* has been included in the book.

Cycads must not be confused with the well-known Tree Ferns *(Cyathea dregei,* now classified as *Alsophila dregei),* which grow along stream banks in the Little Berg, nor with the palm. Both these plants, botanically, are quite unrelated to the Cycad.

Encephalartos ghellinckii grows on the slopes of the Little Berg usually immediately above and below the cave sandstone cliffs. It is found more often in the northern sections of the Natal Drakensberg, between Cathedral Peak and Mont-aux-Sources, than in the more southern areas, although our illustration opposite was taken in the Giant's Castle area. (It is only rarely found in the Giant's Castle area.) There is a large colony of them near the head of the Mhlwazini Valley in the Cathkin area. The plant has a large tessellated trunk, sometimes growing to a height of two to three metres. The leaves are stiff and prickly, with a long sturdy mid-rib. These leaves are the narrowest of all the South African *Encephalartos* species, only about 2,4 mm broad, and this feature readily distinguishes it from all other species in the genus. The plants are dioecious, that is, there are separate male and female plants. The fruit is a characteristic cone, rather like a large pineapple, but without the tuft of leaves at the top. The male cone is greyish in colour, and the female brown and slightly woolly. The seeds have the shape and size of a date.

The word Encephalartos comes from two Greek words meaning "bread within the head", and its common name is the Bread Fruit. Thunberg, the Father of South African botany, described once how the Hottentots scooped out the starchy pith from the trunk of a species of *Encephalartos,* buried this for two months in the earth, and then kneaded it into cakes which they baked in the embers.

Most species of *Encephalartos,* not the leaves or stems, but the seeds, are poisonous. These have a fleshy, outer pulp and a hard-coated kernel. It would appear that the outer pulp is not poisonous, but the hard kernel is extremely poisonous. Baboons and many fruit-eating birds relish the outer pulp, but they always discard or regurgitate the kernel. African deaths have been reported among Pondos who ate seeds of *E. villosus.* In 1912 two Pondo boys died after eating the roasted kernels of an *Encephalartos* species. They began vomiting six hours after eating it. The vomiting then ceased but the boys lay paralysed, unable to rise, and died after 80 hours. According to Medley Wood the Zulus in earlier times ate the seeds, but these had first to be steeped in water, otherwise they were poisonous.

Our Drakensberg species was named after M.Ed. de Ghellinck de Wolle, a noted amateur horticulturist from Ghent.

Cycads are interesting plants, for, unlike most of the rest of our flora, they have come down to us, virtually unchanged, from the very earliest times. They are sometimes called living fossils, and their ancestors were widespread on earth 200 million years ago, long before man appeared only two to three million years ago. What has enabled them to survive for so long is something of a mystery. To-day there are a few species of Encephalartos in Tropical Africa, and about twenty in South Africa as a whole. The genus is found nowhere else in the world. *E. ghellinckii* is the only species found in the Drakensberg.

The first picture, opposite, is the female cone of *E. ghellinckii,* showing the seeds. The cone has been broken down, probably by baboons anxious to get at the seeds. The picture, together with that of the male cone, was taken in the Solitude area of the Drakensberg.

1

2

3

1. *Encephalartos ghellinckii,* female cone showing seeds.
2. *Encephalartos ghellinckii,* photographed in the Giant's Castle area.
3. The male cone of *Encephalartos ghellinckii*.

High up on the summit of the Drakensberg, among the mists and the lonely peaks of Lesotho's Sehlabathebe Mountain National Park, an exciting discovery was made in 1970. Kate Williamson, a young British botanist, was visiting the Park, and floating on the surface of one of the clear rock pools in the cave sandstone she found the white stars of what she was sure was a new species of aquatic plant. She collected specimens and sent them to Kew, and they were indeed pronounced new to science. They were given the name *Aponogeton ranunculiflorus*.

As far as is known, the distribution of these new and lovely flowers, which have been called "the crown jewels of Sehlabathebe", is confined to a narrow strip, 10 km long, in the Park, at 2 600 m, but stretching down to a few pools on the Natal side of the Drakensberg. They are found nowhere else in the world.

The plants grow in the soft, humic mud of the pools, in the clearest of water, from 200 mm to 3 metres deep. The leaves are from 70 to 100 mm in length, are few in number, and always grow under the water. The flower head is borne at the top of a long, slender peduncle (stem) and the flowers rest on the surface of the water. In appearance they look rather like white buttercups (hence the specific name), about 10 mm in diameter, a clear sparkling white. The stamens are bright yellow.

The flowers of this lovely species, floating peacefully on the blue waters of these mountain pools amidst the silence of the great peaks, are indeed one of the treasures of Sethlabathebe.

Another member of this family is the well-known and much more widely-spread *Aponogeton junceus*, (the Water-uintjie). This species is found throughout South Africa wherever vleis occur. It usually flowers from November to January. The inflorescence is a two-branched spike, about 10 mm across and the flowers are bisexual (i.e. with both stamens and pistil). The leaves are long and thin and the flowers rise above the water.

All the plants in this family are water-plants. The name Aponogeton is a partial anagram of *Potamogeton,* another aquatic plant. The word Potamogeton itself comes from two Greek words, *potamos,* meaning a river and *geiton,* a neighbour.

1

2

3

1. ''The crown jewels of Sehlabathebe.'' The white buttercup-like flowers of *Aponogeton ranunculiflorus*.
2. Growing in the humic mud at the bottom of the pool, the flowers float on the surface of the water.
3. Rock pool at Sehlabathebe, with *Aponogeton ranunculiflorus*.
4. *Aponogeton junceus*.

4

Cyperus obtusiflorus var. flavissimus (it used to be called *Cyperus compactus)* is a small plant, only about 100 mm high, bearing bright yellow inflorescences on stiff, slender stalks. These inflorescences form compact, round heads, and what looks like petals are really a series of stiff, brittle, bright-yellow bracts. *Cyperus obtusiflorus* is one of the sedge-like plants, and is usually found in damp places, on the sides of vleis, at about the 1 850 metre level. It flowers in December and January.

The genus is widespread throughout the world, in all warm and temperate regions. The papyrus of the ancient Egyptians *(Cyperus papyrus)* is a member of the genus, which is allied to the grasses. Many species which used to be classified under *Cyperus* have been transferred to other genera. Even so there are something like 550 species altogether. As far as I know, the Drakensberg is home to at least five of them.

The generic name comes from the Greek word *kupeiros,* meaning sedge, and *obtusiflorus* comes from two Latin words, *obtusus* meaning blunt or dull, and *flos,* a flower.

ARACEAE

The *Araceae* family has about 115 genera and over 2 000 species. Although a few genera appear in temperate regions, it is essentially a tropical family, being found on both sides of the equator. What appears to be a large petal is a spathe (really a modified leaf). This spathe is usually waxy in texture, and is sometimes brightly coloured. It encloses a long, usually yellow cylindrical spike, called a spadix. This is actually made up of a mass of very small flowers, tightly packed together. Our two well-known house-plants, Anthurium and Philodendron, belong to this family, and of course there is the well-known Arum Lily.

ZANTEDESCHIA *Araceae*

The correct name for our Arum Lily is Zantedeschia. The genus was originally named *Calla,* the Greek word for 'beautiful', by Linnaeus. Later this was changed to *Richardia,* after a Frenchman, and this in turn to the presently accepted botanical name, *Zantedeschia*. It was named after an Italian botanist, Giovanni Zantedeschi (1773-1846). Common names are Varkblom, Pig Lily, Trumpet Lily, and of course Arum Lily. Pig Lily is a strange name for so lovely a flower. It was given because pigs are particularly partial to its tubers, and this is unfortunate. Dorothea Fairbridge, in particular, disliked it. ("freak of nomenclature", she called it). Conrad Lighton, the well-known authority on the flowers of the south-western Cape, also enters the fray, stating his preference for the name "Calla". "Pigs or no pigs", he says in his delightful book "Cape Floral Kingdom", "what could be more euphonious than the Calla Lilies of the Cape?'

The plants usually grow to a height of nearly a metre, but in September 1956, according to Conrad Lighton, the Cape Argus published a picture of an Arum with a 1,8 m (6 ft.) stem. It was found in the Franschoek Valley.

Arum 'flowers' are large, and consist, first of all, of the usual cylindrical spike, the spadix. This is almost always bright yellow, about 60 mm long, and consists of the true flowers, male near the tip and female near the base. The spadix is half enclosed by the spathe. This is actually a sheathing leaf, but it is this that gives the inflorescence its striking appearance. This spathe is usually a glistening, pure white, but in some species it is a golden yellow, lemon yellow, and even pink. These 'flowers' grow at the tips of thick, green, fleshy stems, well above the leaves, which grow in a tuft at the base of the plant. The leaves vary in shape according to the species, but all are pointed. Some are flecked with white.

There has, in the past, been considerable confusion in the matter of naming the different species. Not only has the generic name been changed twice, but the names of many of the different species have also been changed. To add to the confusion, the plant hybridizes easily, and it also changes materially according to its habitat.

At the time of writing the Natal species have been narrowed down to three, *Zantedeschia aethiopica, Z. albomaculata,* which is divided into three sub-species, and *Z. rehmannii*. In the Drakensberg we probably

1

2

3

1. *Cyperus obtusiflorus* var. *flavissimus*.
2 & 3. *Zantedeschia albomaculata* subsp. *albomaculata*.

have only *Z. aethiopica* and *Z. albomaculata subsp. albomaculata*.

Zantedeschia aethiopica is the well-known white species with clear green leaves, so common in our gardens. Its old names, *Richardia africana* and *Calla aethiopica* have now been discarded. The spathe is pure white, the spadix yellow, and the leaves deep green and shaped like an arrow. They are roughly twice as long as they are broad. The Zulu name is in-Tebe emhlophe. The Zulus heat the leaf and apply it to sores and insect bites as a poultice, and the young leaves are sometimes cooked and eaten as a vegetable. *Aethiopica* means of Africa.

Zantedeschia albomaculata subsp. albomaculata is a little smaller than *Z. aethiopica*. It includes the old *Z. oculata, Z. hastata* and *Z. melanoleuca*. The flowers, too, are smaller than *Z. aethiopica*. The spathe is creamy-white, with a dark purplish blotch on the inside at the base of the spathe. The spathe often shades to a yellowish-green at the base. The pointed leaves are flecked with white and are arrow-shaped. This species grows at somewhat higher altitudes to the last.

The Amangwane, in the foothills of the Drakensberg, prepare a poultice from the fresh green leaves of *Zantedeschia albomaculata* for the breasts of nursing mothers. The tubers (as with *Z. aethiopica),* are often eaten as a vegetable.

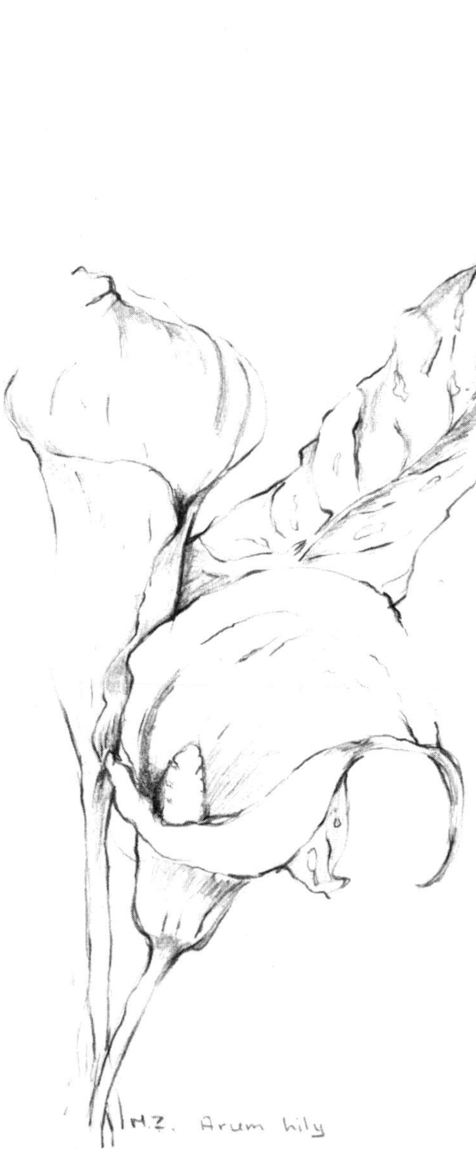

Zantedeschia aethiopica

COMMELINACEAE

The family of the Commelinaceae was named after J. Commelin, a Dutch botanist. It is mainly a tropical family, and is found in most warm countries. Plants are usually low-growing and somewhat sprawling. The family is characterised by flowers that have three small sepals and three blue, yellow or white petals. Altogether there are about 40 genera and 500 species.

CYANOTIS

Commelinaceae

Cyanotis foecunda is a charming and rather unusual little flower. (Up to recently it was called *Cyanotis speciosa,* and before that *Cyanotis nodiflora).* The flowers are a brilliant purplish-magenta and consist of three petals which join to make a cylindrical tube. There are six yellow stamens, and the filaments are bearded with beaded hairs, giving the whole inflorescence a fluffy appearance. Leaves and stems are hairy, and the flowers are borne in the axils of the leaves. It flowers in January and February.

Common names are Doll's Powder-puff and Jacob's Tears. The roots are used medicinally by the Xhosas, and also as a love charm.

The genus as a whole consists of about 50 species, and occurs in all the warmer regions of the Old World, from Africa to South China and north Australia. In South Africa we have only four species. *C. foecunda* is widespread, occurring in the southern and eastern Cape, Transkei, Lesotho, Natal, Transvaal, Botswana and on into tropical Africa. In Natal it occurs all the way from the coast to the Drakensberg.

The word *Cyanotis* comes from the Greek word *kuanos,* a dark-blue mineral, while *foecunda* comes from the Latin word *fecundus (foecundus),* meaning fruitful, fertile, abundant.

COMMELINA

Commelinaceae

Species of *Commelina* are dainty little plants, low-growing, small, spreading and bearing yellow or blue, rather fragile flowers, with broad petals. The flowers protrude from a folded bract or funnel-shaped spathe, and they fade almost as soon as they are picked.

There are at least two Drakensberg species, *Commelina africana* with yellow flowers, common throughout Africa, and *C. erecta* with blue flowers. In the latter the stems, unlike the other Commelinas, stand erect. The common name for the flower is the Wandering Jew.

The Zulu name for the plant is indangabane, (I have also been given ishlambezwa), and the Sotho name lekxotswana. The plant is used by the Zulus as a pig food, and also as a cure for female sterility. The root when crushed and boiled is sometimes used to soothe pain and is also supposed to assist women in child-birth.

Altogether there are 12 species of *Commelina* in South Africa. The genus is widespread throughout the world in warm climates.

1. *Cyanotis foecunda*
2. *Commelina africana*

1

2

LILIACEAE

This is one of the largest families of the Monocotyledons (it has something like 250 genera, nearly 4 000 species, and is distributed world-wide). It is well represented in South Africa especially in the Drakensberg. *Liliaceae* are distinguished from the closely related *Amaryllidaceae* in that they have six stamens and a superior ovary (i.e. the petals are inserted below the ovary), while the *Amaryllidaceae* have six stamens and an inferior ovary (petals inserted above the ovary). The majority of these plants have bulbs, corms or tubers. The leaves are usually long and tapering, and all have parallel veins. Our well-known Aloes form one of the 250 genera.

ALOE

Liliaceae

Aloes are usually regarded as drought-resistant plants, and it would seem surprising that at least eight, and possibly nine, species are known to occur in the Drakensberg, where the rainfall can be as high as 2 000 mm per annum. At the same time we must remember that in the Drakensberg, despite the high rainfall and mists, there are many habitats with shallow soils over impervious rock where the rainwater drains away rapidly and soil dries out within a few hours. Plants in such situations must survive the drought periods between rains, and it is in places like this that most of the Berg Aloes occur. But aloes do grow in a wide variety of habitat, and some actually thrive in rainy conditions.

There are about 350 described species of *Aloe,* occurring mainly in Africa, but they are also found in Arabia, Madagascar, and the neighbouring islands, and also to a lesser extent in India, China and the Mediterranean region. The generic name *Aloe* was given by Linnaeus (1707-1778). There is some doubt as to the actual derivation of the name. It is either derived from the Arabic name for the plant, *alloeh,* or from an old Greek name, allied to the Hebrew *allal,* meaning bitter, referring to the bitter juice of the leaves of all plants in this genus. It is possible that philologically these two words may be connected.

There are 137 known species that occur in South Africa. Eight (possibly nine) of these occur in the Drakensberg.

Perhaps the commonest is *Aloe saponaria.* This is a medium-sized Aloe, about 0,5 m high, growing on the lower slopes of the Little Berg and on rocky outcrops. It flowers throughout August and September, and has a wide range of colour, from yellow, and through orange and pink to a bright brick-red. It belongs to a large group of Aloes known as the Maculates (spotted). The leaves are white-spotted. The name 'saponaria' means soapy, (from the Latin *sapo,* meaning soap. C/f our saponaceous, soapy). One of its common names is the Soapy Aloe. Another common name is Bontaalwyn, while the Zulu name for it is i-Cena. It is one of the most widely-distributed of our Natal Aloes, occurring from close to the sea right up to the Drakensberg. This Aloe has great resistance to the destruction caused by grass fires, only the tips of the leaves being burnt off and killed. The bases of the leaves are so tightly packed that they protect the stems from the flames. It is claimed that the leaves, split open and applied to sores or ringworm, are an excellent cure. *Aloe saponaria* is used extensively by Zulus, Xhosas, Sothos and even Europeans for medicinal purposes, while the Southern Sotho use it as a charm against lightning.

Aloe aristata is a small Aloe, only about 150 mm in height, and usually grows in fairly moist situations below the sandtone cliffs, on the edges of forest areas and on moss-covered rocks in the forests. But elsewhere in South Africa it is found growing in semi-arid conditions. It usually grows in dense, compact groups, the leaves forming a rosette from 100 to 150 mm in diameter. It flowers throughout November and December. The flowers are dull red above and creamy-pink beneath. The Sesutho name for this Aloe is serelei, the Slippery One. In Lesotho an uprooted plant is placed on a shelf in the hut of a barren woman. If it flowers under these conditions, it is believed, she will become pregnant, but she will remain barren if it withers. The botanical name *aristata* means bristly.

Aloe boylei is sometimes mistaken for a lily. It is actually a medium-sized Aloe, about 0,5 m in height and grows on the slopes of the Little Berg in open grasslands, at a height of about 1 850 m, although the picture on page 33, No. 3 was taken on a blustery January day amongst the basalt rocks on the lower slopes of Cathkin at an altitude of 2 150 m. It flowers from November to January. It is a typical mountain Aloe, and likes a high rainfall. It has salmon-pink flowers. The Zulus call it isi-Phuthumane, or isi-Phukuthwana. They eat the young flower-shoots as a

1

2

3

1. *Aloe saponaria* in the Mnweni valley. Note how only the tips of the leaves have been scorched by the fire.
2. *Aloe aristata*.
3. *Aloe saponaria*.

vegetable, and they also make a drink from the sap which is supposed to help in child-birth.

Aloe kraussii is a fairly rare Aloe in the Drakensberg, and is only from 300 to 400 mm high, one of the smallest of our Natal Aloes. It grows in damp places, and the flowers are lemon-yellow in colour and pendulous, appearing in December. Like *Aloe boylei* the young flower-shoots are eaten by the Zulus as a vegetable. The Zulu name is isi-Phukuthwana, and the Sesutho name Lekhalana. Young girls undergoing initiation rites have to bathe themselves on the first night of the initiation with a lotion prepared from this species. A decoction of the roots is drunk by barren women to ensure pregnancy.

Aloe dominella, like *Aloe boylei,* is another true mountain Aloe, the plants being found in hilly country, on rocky hill slopes. Plants are of average size (about 0,5 m high), and grow in groups, as the picture opposite shows. The flowers are yellow, and they bloom from June to December. Strangely enough, they may not flower in cultivation, but G. W. Reynolds, the South African authority on Aloes, claims that this can be overcome by packing dry grass round the plant, setting it alight, and giving the plant a good watering afterwards.

Aloe cooperi is easily recognised from its long, slender graceful leaves, which spring from one point in the ground and are somewhat fan-shaped. They arch over, and have a well-defined midrib, or keel. The lower surfaces, especially near the base, are spotted. The flower-stalks, also, are long and slender, and in the Drakensberg the plants are about 0,5 m high. Along the coast more robust forms are found, and in the northern coast regions of Natal heights of up to 1,5 metres have been recorded. It is often found wedged between rocks and on rocky hill-sides. It flowers in January and February. The colour of the flowers is usually salmon-pink, but is sometimes apricot to deep red. They are sometimes tipped with green. The Zulu name is the same as for *Aloe boylei* and *A. kraussii,* and again, the young shoots and flowers are cooked and eaten as a vegetable. The Zulus believe that if the smoke from a burning leaf of *Aloe cooperi* is allowed to pervade the cattle kraal it will protect the cattle from the ill effects of eating improper food. This dainty Aloe is well worth a place in any garden.

Aloe arborescens is one of the best-known South African Aloes. It grows sparsely in the lower reaches of the Little Berg. The name means tree-like. It is widespread, and is found in mountainous regions from the Cape through to Rhodesia and Malawi. It usually grows in large clumps in rocky places, and the flowers range in colour from coral-red to scarlet, orange and yellow-green. It is frequently used as a hedge round kraals, especially in the Transkei. There is a magnificent one round the camp site in the Hluhluwe Game Reserve. It grew in the Dutch East India Company's garden at the Cape in 1695. Its common name is Kransaalwyn, while the Zulu name for it is in-Kalane, or sometimes in-Hlaba. It is a well-known fact that the leaves are most useful in the treating of X-ray burns. A decoction of the leaves is used by the Zulus to assist in child-birth. It flowers during June and July.

I had to work particularly hard to get the picture of *Aloe pratensis* on the next page. For years I had hunted this Aloe without success, and then, one April, I found a small colony near the summit of the Little Berg in the Cathkin area. It meant a morning's climb to reach the spot and get back. I knew the flowering time was September and October, so five months later, during the first week in September, I paid my first visit. The buds were just beginning to show, and it meant four more hard climbs, keeping constant watch on the development of the flowers, before I was able to get my picture, towards the end of October.

Aloe pratensis is a small Aloe, only about 0,4 m high, short and stocky, with thick, fleshy leaves. It usually grows in basalt outcrops, along the top of the Little Berg, and is another true mountain Aloe. It is often found wedged between rocks on sloping ground.

The flowers are rose-red to orange in colour, and they open during September and October. As the bud develops, the unopened buds and bracts form a tight cone at the top of the stem, and this gradually opens out into the full flower, a feature which is very typical of this Aloe, and which makes it easily recognisable.

Although a mountain Aloe, the name 'pratensis' means growing in meadows, a rather strange name for such an Aloe. Barbara Jeppe, in her excellent book "South African Aloes" says that this suggests rather faulty observation on the part of its discoverer. It was J. G. Baker who named the plant after seeing it growing in W. Wilson Saunders' garden at Reigate, Surrey. But he had probably never seen the plant growing in its natural habitat. In any case, Saunders received the material in 1858, and it was not until 1878 that Baker

---

1 & 2. *Aloe cooperi.*
3. *Aloe boylei,* on the slopes of Cathkin Peak.
4. *Aloe dominella*
5. *Aloe boylei.*

1

2

3

4

5

saw the plant. The original collection records could easily have been lost in those 20 years.

The Basotho name for the Aloe is Lekhala qhalene, meaning "The Aloe which opens out", referring, no doubt, to the developing cone of buds and bracts mentioned above.

There is one other possible Drakensberg Aloe, but this is a little doubtful. This is *Aloe striatula*. Reynolds reports it as growing along the tops of mountains in the Eastern Cape and "also reported from Teyateyaneng, Lesotho". Barbara Jeppe reports it as occurring in southern Lesotho. So far it has not been recognised as a Drakensberg plant, but it certainly grows at Pine Outpost, below Giant's Castle. Mr Eric Jensen, Curator of the Natal Parks Board Garden at Giant's Castle, feels, however, that it may be an introduced species.

Widespread as it is, and so typically South African, the Aloe could well have been South Africa's national flower. In fact, at the beginning of the century, an attempt was actually made to achieve this. In 1904 Dr F. E. Ensor published a pamphlet "The Aloe: An Emblem and its Meaning" advocating just this. Unfortunately for the Aloe, the Protea won the day!

---

1. *Aloe arborescens.* The picture was taken on the slopes of the Little Berg in the Cathedral area, with the Cathedral Range in the background.
2. *Aloe pratensis,* showing cone of unopened buds and bracts.
3. *Aloe aristata.*

1

2

3

ALBUCA

Liliaceae

Hidden in the long grass of the Little Berg you will sometimes come across a charming little flower rather like a snow-drop. The petals are off-white, with a green stripe running down the centre, and tinged with a golden- or purplish-brown. This is *Albuca setosa*. (Until recently it was called *Albuca pachychlamys*). Its common name is Soldier-in-the-box, the name referring to the inner stamens which stand against the upright (inner) perianth segments like a sentry in his box.

Albuca flowers are easily recognised by their shape. There are six petals. The three inner petals curve inwards, almost forming a tube, while the three outer petals swing outwards. The plant has a fairly large bulb with thick tunics or scales which are attached by a series of wiry fibres. The word Albuca comes from the Latin word *albus,* meaning white, for most of the species of this genus are white in colour. The leaves are short and narrow. Unlike the other Albucas, whose flowers droop downwards, the flowers of *Albuca setosa* stand erect. The plant flowers throughout early spring and summer.

Altogether there are about 100 species of *Albuca,* mainly from Africa. A few species come from Arabia, but the plants are found nowhere else. South Africa can lay claim to about 50 species, and these extend from South-West Africa across to Natal, where there are seven species. Some are small plants, only about 150 mm high, but a few reach a height of a metre or more. *Albuca setosa* is only about 200 mm high. At least three species of *Albuca* are to be found in the Drakensberg.

The Zulu name for the flower is i-Chiye. The Zulus crush the bulbs, add a little water, and then proceed to wash their oxhide shields in the mixture. This helps to bring out the rich colour and sheen of the ox-hide. *Albuca tricophylla* is used by the Sothos in the treatment of gonorrhoea.

LEDEBOURIA

Liliaceae

The genus *Ledebouria* contains a number of species previously classified under *Scilla.* In Natal we have nine species, eight of which used to be included in the genus *Scilla.* Three species at least occur in the Drakensberg, *Ledebouria cooperi, L. ovatifolia,* and *L. revoluta,* the latter with erect, narrow leaves and a large bulb. It is possible that *L. luteola* also occurs, though I have not come across it myself.

Ledebouria ovatifolia is a small plant, only about 100 mm high. A single slender stem arises from two egg-shaped leaves (hence the specific name), which lie flat on the ground. They are slightly speckled. The flowers are small, and grouped in a short spike of 50-150 densely packed flowers. They are pinky-red in colour. The plant is often found growing on the dry soil of gullies in the Little Berg at an altitude of about 1 500 m, and it flowers in January and February.

Ledebouria cooperi is larger than *L. ovatifolia,* reaching a height of 200 mm. Leaves are erect, with distinct parallel veins. This species is found in moist situations, at about the same altitude as *L. ovatifolia.* The flowers are white to pinkish, and they appear from October to December.

1

2

3

4

5

1. *Ledebouria revoluta*
2. & 3. *Albuca setosa*
4. & 5. *Ledebouria ovatifolia*

Ever since they were first discovered in South Africa, well over 200 years ago, these strange plants have intrigued horticulturists. The flowers are easily recognised, for they are massed together on a perfectly cylindrical spike, and on the top of the spike is a crown of leafy bracts. It reminds one of a pineapple, and Pineapple Flower is actually one of its common names. Children in Natal sometimes call the flowers Ricksha-boys, obviously on account of the 'top-knot', reminiscent of the showy plumes of the ricksha-boys. The botanical name, *Eucomis,* also refers to this tuft of bracts at the top of the spike, for it means beautiful hair, or top-knot. The word comes from the Greek. The flower-spike rises from the centre of a large rosette of dark green leaves, oblong in shape. The individual flowers are all about 10 mm in diameter.

The plants are usually found growing on moist inclines, and in grassy gullies at about the 1 800 metre level. They have a long flowering period, from December to March, and they are from 250 to 300 mm high.

Eucomis is essentially an African genus. There are 10 species altogether, one of which is found in tropical Africa. The rest all come from South Africa, with seven from Natal, three of which occur in the Drakensberg.

Eucomis bicolor is the showiest of the three. The flowers are pale red to mauve, and all have dark purple centres. It was first cultivated in England about 80 years ago. It is used by the Southern Sotho as a remedy for colic.

Eucomis humilis is similar in many ways to *Eucomis bicolor,* but the flowers are all green. The leaves of this species are spotted. *Humilis* is a Latin word meaning small, low or slight. Our word humility comes from it.

Eucomis autumnalis var. autumnalis is the new name for the old *Eucomis undulata.* Here the flowers are pale green to whitish, very similar to *E. humilis,* but in this species the leaves are unspotted. They are also wavy on the margins, hence the old name. The flowers are sweetly scented. According to records this species was growing in the gardens of Europe over 200 years ago. It is possible that seed was sent over from the Sneeuberg or Uitenhage districts. In 1760 P. Miller used seed from the Cape to raise plants in the Chelsea Gardens.

The Zulu name for *Eucomis autumnalis* is u-Makhonde, and it is believed by the Zulus to be a powerful charm against witchcraft. They also use it in the treatment of urinary diseases. The Tswanas boil it with water, milk or meat and use the mixture for the relief of abdominal pains. Transvaal Sothos use the bulb as an effective remedy for the inevitable hangover after a 'night out with the boys'!

1. *Eucomis bicolor*
2. *Eucomis autumnalis var. autumnalis*
3. *Eucomis humilis*

1

2

3

There are ten species of *Agapanthus,* all endemic to South Africa, occurring from the southern Cape to the eastern regions. They are all strikingly beautiful plants, spreading by means of rhizomes which produce a multitude of dark green, strap-shaped leaves, and large umbels of bright blue or pale blue flowers. The flowering stems are fairly stout and may reach a height of a metre or more.

The name means 'dearly-loved flower', and comes from two Greek words *agape,* meaning love and *athos,* a flower. It is found wide-spread throughout South Africa, but strangely enough is found nowhere else in the world in its wild state. It was introduced as early as 1687 into cultivation in Europe, and Linnaeus knew it. It is known that specimens of *Agapanthus* flowered in the Royal Gardens at Hampton Court in England in 1692, during the reign of William of Orange. Since then the plants have flourished, especially in England, often surviving up to 30 degrees of frost. According to Conrad Lighton they have spread to St. Austell, on the Cornish coast, where the cliffs are smothered with them, "their blue competing with the blues of the Cornish sky and the English Channel" according to a London magazine. They have done well even as far north as the north-west coast of Scotland where, year after year, Mr Osgood McKenzie's waterside garden at Inverewe astounded horticulturists with its magnificent display of Agapanthus. In the Drakensberg they can be seen in their full glory, growing on dripping wet ledges in the Little Berg.

The flower was first named by Linnaeus in 1753, and he called it *Crinum africanum.* Later the genus *Agapanthus* was established. One of the Drakensberg species is *Agapanthus inapertus.* Four other species are known in Natal, but there are numerus sub-species and varieties. Common names are African Lily, Blue Lily, Bloulelie, and Haaklelie. The Zulu name is u(lu)-Bane. The Zulus make an infusion from the roots to sprinkle their yards and huts with, as a protection against storms. It is also used to *phalaza* with by a person who is afraid of thunderstorms.* Sometimes a lotion is made from the flowers to wash newly-born babies, in the belief that it will make them strong. Xhosas use the roots to make a paste for the treatment of swollen legs.

*To 'phalaza' is one of the oldest customs of the Zulus. An infusion is made from certain herbs or roots, and mixed with hot or cold water. The Zulu first drinks this. He then takes up a kneeling position and tickles his throat to cause vomiting. This is supposed to cleanse the stomach and act as a charm.

1

3

2

1 & 2. *Agapanthus inapertus*
3. *Agapanthus sp.*, probably *inapertus*.

The *Kniphofia* genus was named after Johann Hieronymus Kniphof (1704-1763), Professor of Medicine at Erfurt University in the latter part of the 18th century, but the flowers are better known as Red Hot Pokers. They make a very attractive garden plant. The majority of Kniphofias flower at the begining of autumn, and their brilliant red and orange spires against the blue of the mountains is the sign to the mountaineer that the blue-gold days of autumn are upon him.

The genus is essentially an African one, with one outlier in Madagascar. It is a complicated genus and a good deal of work still has to be done on it. Many of the early specific names have now been abandoned, and new ones applied. There are quite a number of different species in the Drakensberg, and it is sometimes a little difficult to differentiate between them.

The flowers are tubular and arranged in a tapering spike at the top of a firm, erect peduncle, or stalk. The flowers commence opening at the base of the spike while those at the top are still in bud. The throats of the flowers are usually long and narrow, and I have often found that bees, forcing their way into the flowers, are caught and are unable to return.

Kniphofias are closely related to the Aloe, and some species are sometimes confused with Aloes. Up to 1769 the plant was actually classified as an Aloe, but in that year it was placed in a separate genus. If ever you are in doubt, remember that the essential difference between the two genera is that Aloes have thick, fleshy leaves while Kniphofias do not.

Kniphofia caulescens is a tall plant, growing to a height of a metre and more, found on the summit of the Drakensberg, often on the edges of small pools and vleis, their brilliant orange spikes reflected in the clear water. Flowering period is December and January. This species is easily recognised by its leaves. These have a ridge, like the keel of a boat, down their centres, and the leaves are broad and long. The word caulescens comes from the Latin word *caulis,* meaning the stalk or stem of a plant (our word cauliflower also comes from the same word) and means with a clearly visible stem.

Kniphofia breviflora is a small plant, as the name implies. It only grows to a height of about 200 mm, and is usually found in grasslands at an altitude of about 1 500 m. It flowers from January to March, and the flowers are a bright yellow, although there is also a scarlet variety. Unlike most of the Kniphofias, the stems tend to curl downwards at the tips. The photograph opposite was taken in the Setene Valley during a never-to-be-forgotten trip in April 1959.

Kniphofia ichopensis is a tall, slender plant, up to a metre in height, growing in grasslands at about the 1 800 m level. It flowers in midsummer, and the flowers vary in colour from a creamy-white to a coral pink. The leaves are long and slender.

One of the most colourful of the Kniphofias is *Kniphofia evansii.* With its brilliant orange flowers it is indeed an eyecatching sight in the grassy gullies of the Drakensberg, at a height of about 2 000 m.

We might briefly notice four other Drakensberg Kniphofias. *Kniphofia rufa* is long and slender, about 300 to 400 mm high, bearing spikes of yellow and red flowers and thin, slender leaves. *K. ritualis* is small, with broad leaves, about 400 mm high, and it grows usually at the base of vertical cliffs at about the 2 750 m level. It flowers in January. *K. porphyrantha* is a dainty little Kniphofia, with a height of not more than 0,5 m. It has lemon-yellow, tubular flowers, with the buds (at the top of the spike) a pale orange or red. (Actually the word *porphyrantha* means purple flower). It flowers from October to February. *K. multiflora* is the tallest of our Drakensberg Kniphofias. It grows to a height of over two metres, and is also remarkable as being the only Drakensberg Kniphofia which has its flowers erect instead of hanging down. It has a narrow flower spike, 0,5 to 1,0 metre in length, and the flowers are small and very crowded. This has the effect of making it look like a bulrush, and its common name is actually Bulrush Poker. The colour of the flowers is usually white, but varies to yellow with reddish buds. It flowers in autumn, and is often found standing in water. It is found growing between Van Reenen and De Beers Pass and on into Swaziland and the Eastern Transvaal.

There is one other species worth noting, although there is some doubt about it. This is *Kniphofia praecox,* the cultivated species. Some authorities claim that it grows wild, and it certainly has been found in the Drakensberg, but others regard it as an escape from cultivation. J. H. Ross does not list it in his authoritative "Flora of Natal". It forms thick clumps, and the flower

---

1. *Kniphofia caulescens,* growing beside a mountain pool on the summit.
2. *Kniphofia evansii.*
3. *Kniphofia rufa.*
4. *Kniphofia breviflora.*
5. *Kniphofia ichopensis.*

1

2

3

4

5

stems reach to over a metre in height. It flowers during winter.

The Zulu name for all Kniphofias is u-Mathunga.

ORNITHOGALUM

Liliaceae

The Genus *Ornithogalum* is essentially a mountain genus, though a few species are found at lower levels. The well-known Chincherinchees, from the Cape, belong to this genus. Fourteen species are to be found in Natal, and at least four of these are known to occur in the Drakensberg.

The flowers are almost always pure white, and rather like a snow-drop, but there are a few yellow or orange species. The plant usually grows to a height of about 250 mm, and the six-pointed 'petals' are borne at the tip of a usually stiff, erect stem, though one of the pictures opposite indicates that sometimes this stem arches over. The leaves form a tuft at the base of the plant. Although it is found in various parts of South Africa, it is essentially a Cape flower.

Ornithogalum paludosum is a slender white flower, about 600 mm high, which grows on the summit of the Little Berg. The flowers, each about 10 mm in diameter, form a long, slender spike, and they usually appear in January.

Ornithogalum diphyllum is a dainty little flower which grows in moist, boggy situations on cave sandstone pavements. It grows to a height of about 80 mm, the flowers are white with yellow stamens, and they appear during December. The photograph opposite was taken at The Lakes, Giant's Castle, at an altitude of about 2 000 m.

The word Ornithogalum comes from two Greek words, meaning birds, and milk, but there is some doubt as to how this is applied to the plant. Some authorities think that this refers to the egg-white tint of the 'petals', but others are of the opinion that the word simply means birds' milk, and that this is a Greek proverbial expression for something exceedingly precious or dainty.

1

2

3

4

5

6

1. *Kniphofia porphyrantha.*
2. *Kniphofia ritualis.*
3. *Ornithogalum sp.*
4. *Ornithogalum diphyllum.*
5. *Kniphofia praecox.*
6. *Ornithogalum paludosum.*

One of the spring flowers you are not likely to forget is the blue Scilla, its delicate plumes waving gently in the air of a sunlit morning.

The genus used to be a large one. Professor J. W. Bews, in 1921, was able to list no less than 26 species occurring in Natal. Since then most of these have been transferred to other genera, such as *Ledebouria* and *Drimiopsis,* and to-day, in Natal at any rate, only two species are recognised, *Scilla natalensis* and *S. nervosa,* with two more in the rest of South Africa, making a total of four.

Scilla natalensis, a relative of the English Hyacinth, is one of the commoner flowers to be found in the Drakensberg in early spring. It consists of a mass of tiny pale-blue flowers borne on a raceme at the tip of a long, slender stem, which often reaches a height of 0,8 m. The flowers start blooming at the base of the raceme proceeding to the tip, which often droops over in a gentle curve. The effect is that of a graceful plume of misty blue. The flowering stem rises from a large, rounded, tunicated bulb, which often grows more than half out of the ground. (How often has the tired climber blessed this tough, deep-rooted projection as he struggles up a steep eighty-degree slope!) The leaves are broad and tapering, and emerge in a sort of tuft from the top of the bulb. The plant grows in the Little Berg, at an altitude of from 1 400 metres to 1 800 metres, below the sandstone rock-bands, often in the sides of gullies and water-courses. It flowers in October.

Common names are Blue Scilla, Blue Hyacinth, Blue Squill and Blouslangkop. There are two Zulu names for it, ubulika and in-Guduza. The bulbs are often eaten with food as an aperient by the Zulus, who also cut up the bulbs and use them as soap.

The plant is poisonous to stock. It causes lung-sickness in cattle, and sheep who eat it are known to die within a few hours. Distribution is from the Eastern Cape, the Transkei, Natal, Swaziland and the Eastern Transvaal.

Scilla nervosa is much smaller, only about 250 mm high. In this case the bulb is completely below the ground. The flowers are greenish-white in colour, and the leaves stiff and rigid, and clearly veined. (The word nervosa comes from the Latin *nervus,* meaning a sinew, tendon or nerve.)

There is another Scilla, very small, about 100 mm high, with dark-blue flowers, which used to be called *Scilla kraussii.* This however has now been included in *Scilla natalensis,* and the name *kraussii* has fallen away.

The derivation of the word Scilla is doubtful. It may come from the Latin word for a sea-onion or sea-leek, *Scilla.* Or, according to some authorities, it may come from a Greek word meaning to injure, because the bulbs contain a virulent poison.

The Southern Sotho use the powdered bulbs of *Scilla natalensis* for sprains and fractures, while the Tswana rub it into their backs and joints in the belief that it will make them strong and resistant to witchcraft.

---

1. *Scilla natalensis.*
2 & 3. *Scilla nervosa.*
4. *Scilla natalensis.*
5. *Scilla natalensis* a close-up of the flowers.

1

2

3

4

5

GALTONIA

Liliaceae

Returning one day in late January down the lovely Mnweni Valley, after ten days on the summit, I came across a moist hollow beside a mountain stream, and I found there a massed display of the lovely white *Galtonia candicans,* the well-known Berg Lily. I had seen it often, of course, in the Drakensberg, but never had I seen such a wonderful display of these beautiful flowers. It was early morning, and the dew was still fresh on the petals in the clear sunlight. It was a sight which will remain long in my memory.

You will find this handsome flower during the month of January. The flowers are pure white, pendulous and bell-shaped, and the plant grows to a height of a metre or more. It dies down completely in the winter, but in the spring it throws up a long stem, and on this the flowers hang down, attached to the main stem by a short stalk. It is a favourite with gardeners throughout the country, for it is easily grown. Its common names are Berg Lily, Summer Hyacinth (like *Scilla natalensis* it is related to the Hyacinth), while the Zulu name is isidwa esimhlope.

The genus is indigenous to South Africa, and found nowhere else, but here it is widespread, occurring in the Eastern Cape, Transkei, Natal, Lesotho and Zululand. It is also found growing wild at Aliwal North, in the Free State, and in the Eastern Transvaal. It is very common in the Drakensberg.

It was named after Sir Francis Galton (1822-1911), an outstanding British scientist, anthropologist, traveller (he travelled extensively in South-West Africa during 1850-1), geographer and statistician. He was also a writer on heredity, and invented the word eugenics. *Candicans* is the Latin word for white, clear and shining.

Galtonia is essentially a mountain genus. There are only two species in the genus, *G. candicans* and *G. viridiflora.*

G. viridiflora (the name comes from the Latin word for green, *viridus)* has green flowers. It grows at a somewhat higher level than *G. candicans,* from 2 200 to 2 500 metres, usually in damp places alongside streams. It is especially common on the sides of waterfalls. Like *G. candicans,* it flowers during January and February.

The bulb of *Galtonia candicans* has a weak digitalis action.

DRIMIOPSIS

Liliaceae

Many of the flowers classified under the genus *Drimiopsis* used to be classified under *Scilla. Drimiopsis lachenalioides* is one of them. It used to be called *Scilla lachenalioides.* It is a small plant, 100 to 120 mm high, found growing on the summit of the Little Berg at about the 2 000 metre level. There are usually two leaves, mauve-spotted, lying flat on the ground, and from these leaves rises a slender stem bearing a head of bright pink, tubular flowers. Flowering time is November and December.

There are five species of *Drimiopsis* in Natal. The word *lachenalioides* possibly comes from the Latin word *lachanisso,* meaning resembling potherbs in character, and therefore weak or feeble. This may refer to the stem, which is soft and flaccid, and usually droops down.

---

1. *Drimiopsis lachenalioides*
2. *Galtonia candicans*
3. *Galtonia viridiflora.*

1

2

3

LITTONIA

Liliaceae

Mairn Hulme, in her "Wild Flowers of Natal", tells of the strange and no doubt embarrassing custom (at least to the girl!) relating to a somewhat rare and dainty little climber, *Littonia modesta,* known to the Zulus as u-Hlamvu lwentombazana. When a young Zulu finds the girl of his choice a trifle unwilling, he collects bulbs of this little creeper and of *Gloriosa superba,* the well-known Flame Lily of Rhodesia, pounds them up and makes a decoction of them. Great care has to be exercised, as the bulbs of both plants are poisonous. He then *phalazas* with this mixture (see footnote page 40). He must then walk slowly past the girl, making sure that his shadow falls on her. In a few days she will look as if she is pregnant, and this condition will persist until she returns his affections. If, however, she still does not want him, her father must take a hand. He must first consult a witchdoctor, an Inyanga. The Inyanga, in his turn, prepares a mixture for the girl to *phalaza* with, using the bulbs of the same two plants, plus a few more roots and bulbs, to give it a bit of extra kick. Her embarrassing condition will then disappear.

Whether the plant possesses these magical properties or not, it is certainly a very lovely flower. The flowers are bright yellow and bell-shaped, the 'petals' are six in number, pointed, and curve outwards. In many ways it is very similar to *Gloriosa superba,* especially in its stem and leaves, but the flowers are very different. It is also very similar to, and often mistaken for, another yellow-flowered creeper, *Sandersonia aurantiaca* (see below). In *Littonia modesta* the bells are open and there are delicate tendrils at the apices of the leaves which cling to surrounding vegetation. In *Sandersonia aurantiaca* the bells curve inwards to a narrow opening, and there are no tendrils at the ends of the leaves.

Common names for *Littonia modesta* are Butter Lily (referring to the yellow colour of the flowers), Christmas Bells and Climbing Bell. Distribution is the Eastern Cape, Transkei, Natal, Swaziland, Orange Free State and the Transvaal. There are seven species to the genus, found only in Africa and Arabia. *Littonia modesta* is the only one that occurs in South Africa.

The name Littonia honours the late Dr. Samuel Litton, Professor of Botany at the Royal Dublin Society for 21 years. *Modesta* is the Latin word for modest, or unassuming, and probably refers to the pendulous flowers.

SANDERSONIA

Liliaceae

Closely allied to *Littonia modesta* is *Sandersonia aurantiaca,* another climbing plant with bright orange-yellow flowers. Each flower is over 25 mm long, and 12 mm in diameter. The perianth is puffed out into the shape of a bell, with the mouth slightly constricted, and the flowers hang down gracefully from a curved, wiry stalk. The leaves are rather like those of *Littonia modesta,* long, slender and tapering. The plant is a little smaller than *Littonia modesta,* 0,3 to 0,45 m high. It is found throughout Natal, Pondoland and most of Zululand. It flowers in December, usually in partial shade, on forest margins.

Common names are Chinese Lantern, Christmas Bells and Geelklokkies. It was named after Mr J. Sanderson, a South African plant collector who discovered the flower. He was one of the Byrne Settlers, who arrived in Natal in 1850, and who established and edited a somewhat controversial newspaper, the "Natal Colonist". *Aurantiaca* comes from the Latin word *aureus,* meaning golden. The Zulus call it i-Hlamvu lasenhla, and they make an infusion from the leaves with which they bathe the body to ward off evil.

The flower appears only rarely in the Drakensberg. For many years I hunted it, unsuccessfully, but finally ran it to earth with the help of two Parks Board officials at the Giant's Castle Game Reserve, Mr Eric Jensen and Mr Paul Miles. The picture opposite is the result of our efforts. *S. aurantiaca* is the only species in the genus, and it appears only in South Africa.

1

2

1. *Littonia modesta*.
2. *Sandersonia aurantiaca*.

There are nine species of *Anthericum* in Natal, but it is actually a fairly large genus, with about 50 species altogether. (It used to be much larger, up to 350, but many of the species have recently been transferred to other genera). They come mainly from Africa but a few are to be found in Madagascar, South America, New Zealand and Japan. At least two are found in the Drakensberg. The flowers in all the different species look very much the same – white, star-like flowers with six petals and very prominent yellow anthers. The main distinguishing features between the different species are the varying arrangements of the leaves and the disposition of the flowers.

The two Drakensberg species are *Anthericum cooperi* and *A. longistylum*. *Anthericum cooperi,* illustrated opposite, is a dainty little plant, about 150 mm high, growing singly in open grassland at a height of about 1 500 metres. It flowers from September to December. The inflorescence is congested at the apex of the stiff, slender stem, where a cluster of white flowers will be found. The perianth segments are marked with a green line down the centre, and the six large yellow anthers are particularly striking. The Basothos call it Lelwana, and they use it as a remedy for hysteria.

AMARYLLIDACEAE

This family is a large one, with something like 85 genera and 1 100 species distributed world-wide, mainly in tropical and sub-tropical regions. It is closely allied to the last family, Liliaceae for the flowers have six perianth segments ('petals') and six stamens, in two whorls. The ovary, however, is inferior, that is, the perianth segments and stamens are inserted above the ovary, and not below as in Liliaceae. Also like Liliaceae (and all Monocotyledons) the leaves are long, narrow and parallel-veined. All the plants in this family, except one, have large bulbs or corms, and the inflorescence is often in the form of an umbel.

Anthericum cooperi

The genus *Haemanthus* is unmistakable. The 'flowers' consist of large flower heads borne at the tips of thick, fleshy stalks, and vary in colour from white to pink, orange and red. The flower-head is surrounded by a mass of coloured bracts. The actual flowers are usually small, tightly packed together, with long filaments, giving the plant the appearance of a shaving brush. The word Haemanthus comes from two Greek words, *haema* meaning blood, and *anthos,* a flower. Hence the common name Blood Flower. Other common names are Torch Lily, Paint Brush, Snake Lily, Varkoor, and April Fool. It is said that the name April Fool was given to the plant because at first it looks as if the inflorescence is a single flower. On closer examination, however, it turns out to be a mass of tiny flowers, tightly packed together. The genus consists of over 90 species, confined almost entirely to the African continent.

Haemanthus magnificus is possibly the best known of all the species. The flower head is borne at the tip of a long, thick, fleshy stalk, which can grow up to 0,6 m in length. The bracts are usually a dark purplish brown, and the flower-head, which at its finest can measure 150 mm across, consists of a mass of very small tightly-packed flowers, the youngest of which are in the centre. The flowers are orange or scarlet, and the long filaments are a brilliant orange-scarlet, tipped with gold anthers. The leaves are bright green and shiny. The pollen is supposed to cause sore eyes, hence another of its common names, Seeroogblom. The seeds are contained in a shiny red, fleshy berry, every bit as spectacular as the flowers. It often grows in the shade of *Podocarpus* forests, in thick leaf mould, and flowers in September and October.

The Zulu name for it is i(li)-Dumbe likanhloyile. The bulb of the plant is ground up by the Zulus, together with various other vegetable ingredients, and this is then boiled, producing a potion called "imbiza". This is used for the relief of gastro-intestinal disorders.

Haemanthus hirsutus is a rather smaller plant. It is characterised by a brilliantly-white flower-head, with numerous pure white stamens and yellow anthers. This gives the flower its typical spiky appearance. It may be found growing in the Little Berg from 1 200 to 1 800 metres, often amongst the sandstone cliffs and on rocky ledges. It flowers during October and November. The bud, a bright pink hemisphere at ground level, is particularly striking. The leaves are bluish-green and hairy, and the stem, also, is hairy, thus giving the specific name *hirsutus,* which is the Latin word for hairy. The Southern Sotho name for the flower is tsebe ea phofu.

Flowers of the genus *Haemanthus* contain poisonous alkaloids in their bulbs, which are supposed to reduce blood pressure. The Hottentots were reputed to have used this poison for the tips of their arrows. There is also an old belief amongst the Hottentots that vultures ate the leaves to sharpen their eyesight.

---

1. *Haemanthus hirsutus*
2 & 3. *Haemanthus magnificus*
4. *Haemanthus hirsutus*. The bud breaks through the blackened veld after a fire.
5. The fruit of *Haemanthus magnificus*.

1

2

3

4

5

One of the first flowers to awaken my interest in the world of wild flowers was the scarlet Fire Lily, *Cyrtanthus contractus*. To this day I can remember wandering as a boy over the burnt veld of my uncle's farm in the Dargle area, and marvelling at their scarlet beauty against the blackened veld long before the rest of that lovely countryside had awakened from the long sleep of winter. To-day, with the march of civilisation, you will have to search long and carefully for them, for they are becoming scarcer and scarcer, but the *Cyrtanthus* genus as a whole still provides some of our loveliest wild flowers, especially in the Drakensberg.

All the plants in this genus have bulbs, and, being long and tubular, the flowers almost always droop downwards. They come in shades of red or scarlet, yellow, orange, white and cream. Altogether there are some 50 species in South Africa. The well-known Ifafa Lily, *Cyrtanthus mackenii,* is one of them. It is essentially a South African genus. Three species are found in Central Africa, but none elsewhere in the world.

I know of at least five species in the Drakensberg. *Cyrtanthus erubescens* is probably one of the loveliest. It is found growing in sheltered places in the Little Berg, and in many of the passes leading to the summit of the Main Berg. Its colour is the most delicate coral-pink, shading to pure white. The specimen shown opposite was photographed in the Mnweni Pass, at an altitude of about 2 250 metres, on a day in January.

Cyrtanthus flanaganii is a yellow species which grows on the summit at an altitude of about 3 000 metres, usually in moist places under basalt outcrops. The flowers are between 30 and 40 mm long, and the leaves are dark green and shiny.

Cyrtanthus tuckii var. viridilobus is only found in the lower reaches of the Drakensberg. The flowers are bright red, with a touch of green at the tips. It is known as the Green-tipped Fire Lily. The name viridilobus comes from the Latin word *viridis,* green and the Greek word *lobos,* meaning a lobe or pod.

Cyrtanthus breviflorus was named originally by Harvey in 1859. In 1878 Baker took it out of *Cyrtanthus,* and created a new genus for it, *Anoiganthus,* calling it *Anoiganthus breviflorus,* and it is as *Anoiganthus* that most of our readers will remember this lovely flower. In 1967, however, largely as a result of the work of Dr K. D. Gordon-Gray of the University of Natal, and Dr F. B. Wright, it was decided that there was no real justification for the retention of the genus *Anoiganthus,* and that the plant should revert to its original classification of *Cyrtanthus*.

The flowers are a bright yellow in colour, about 30 mm long, and are not quite as tubular as most of the other species in this genus. Also the flowers do not hang downwards, as practically all the others do. There are two varieties of this yellow flower to be found in the Drakensberg. One appears on the banks of streams, at the same time as the leaves, in clusters, while the other occurs singly, often on the blackened veld after a fire, and before the leaves appear. Though somewhat dissimilar in habitat and growth, both bear the same name, *Cyrtanthus breviflorus*. Both types appear in the early spring, in the month of August. The Zulu name is u-Velabahleke, meaning the happy, laughing flower – and what a delightful name it is for a delightful flower!

The graceful, drooping flowers of *Cyrtanthus contractus* are one of the first signs of an early spring. The bright flames of their scarlet trumpets appear in the burnt grass of the blackened veld, sometimes even before the first spring rains. This is because the bulbs contain stored water, and the plants are able to shoot even though the ground is still dry. It well deserves its common name of Fire Lily. The Rev. W. C. Holden, one of Natal's early colonists, was one of the first to be charmed by this lovely flower. Writing in his "History of the Colony of Natal" in 1855, he spoke of these Fire Lilies "whose bowed heads seem to supplicate heaven's mercy for the scorched earth, and whose glorious beauty is enough to entice the treasure of the clouds to their feet".

The name Cyrtanthus comes from the Greek words *kyrtos,* meaning curved, and *anthos,* a flower. As we have already seen, nearly all the flowers in this genus curve downwards. The Zulu name for *Cyrtanthus contracus* is u-Mpingizane.

---

1. *Cyrtanthus flanaganii.*
2. *Cyrtanthus breviflorus.*
3. *Cyrtanthus tuckii var. viridilobus.*
4. *Cyrtanthus erubescens.*

1

2

3

4

Apodolirion buchananii is a dainty white flower with a pale pink tube which you will sometimes find, again, on the blackened veld after the grass has been burnt at the end of winter. It grows singly or in small colonies, and the leaves only appear after the petals have died back. There is scarcely any stalk at all, and the flowers lie almost flat on the ground. The flower itself, about 30 mm in diameter, consists of a waxy, satin-like slender tube, fairly long and widening out towards the apex. The six perianth segments, predominantly white but pale pink on the undersides, make it easy to identify the plant.

There are five species of this genus in South Africa, and three of these have been recorded in Natal, *Apodolirion buchananii, A. ettae,* and *A. mackenii.* The common name of *A. buchananii* is the Natal Crocus, but in the Cape it is known as Kukumakranka. The Zulu name is i-Cukudwane (though this may only apply to *A. ettae*). A concoction is made by the Zulus from the bulbs and this is used as a purgative or as an enema for stomach troubles.

Apodolirion buchananii is indeed a lovely little flower, but you will be lucky if you ever come across one, as it is not very common, and the flowers last for only a day or so.

1

2

3

1. *Cyrtanthus contractus.*
2 & 3. *Apodolirion buchananii*

In 1659 a ship from the Far East, sailing to Europe, was wrecked on the shores of the island of Guernsey, after calling in at the Cape. In its hold it carried a large supply of bulbs of a beautiful red lily-like plant called Nerine. The bulbs were washed ashore, took root, found a natural home, and grew well. To this day there is a flourishing export industry on this island in Nerine bulbs. The old Latin name for Guernsey was Sarnia, and so, when this species was brought to the attention of the botanist James Douglas, he named it *Nerine sarniensis,* thinking it was indigenous to the island. Later the story of the shipwreck came out, and it was realised that the plant was not indigenous to Guernsey, but for many years it was thought that the bulbs had come from Japan, where the species had been growing in cultivation for many years. It was not until Francis Masson, the first plant collector to be sent to South Africa from Kew, found the plant growing wild on the slopes of Table Mountain during his second visit to South Africa in 1785-1795, that the true home of the species was finally established. But how, long before this, had the plant reached Japan? Nobody knows.

The Nerine is a most beautiful flower, rather like a rosy-red Agapanthus, the flowers being borne in an umbel at the end of a long, leafless stalk. The six perianth segments ('petals') are re-curved, with long stamens, also six in number, which protrude from the throat. Many of the species glisten and shine in the sunlight, as if they had been spangled with gold-dust.

The genus itself is named after the sea-nymph, Nereides, daughter of Nereus, a sea-god in Greek Mythology who ruled the Aegean Sea. The genus is exclusive to South Africa except for one species which is found in Rhodesia. *Nerine sarniensis* is a Cape flower, and is not found in the Drakensberg, but at least three other species do occur here in these mountain regions.

Nerine appendiculata grows to a height of about 0,8 m and flowers in March and April. It is usually to be found below the sandstone cliffs of the Little Berg, at about the 1 750 metre level.

Very similar to Nerine *appendiculata* is *Nerine angustifolia* both in appearance and habitat. The main difference between the two plants is in the leaves. *N. appendiculata* has strap-shaped leaves, while *N. angustifolia* has narrow, grass-like leaves. The name angustifolia comes from two Latin words, *folium,* a leaf and *angustus,* narrow.

The third Drakensberg species is different from the other two. It grows at a much higher level, being usually found beneath the basalt cliffs on the summit, at about the 3 000 metre level. It is also a much larger plant than the other two, growing to a height of at least a metre. Its flower-head, also, is much larger. Like the other two, however, it flowers in autumn. I once found a magnificent specimen nearly 1,25 metres high, growing below the high basalt cliffs between Sentinel Peak, in the Royal National Park area, and the Chain Ladder. Its name is *Nerine bowdenii.*

The Southern Sotho name for *Nerine angustifolia* is lematlaa. In Lesotho the dry bulb, mixed with parts of other plants, is used as a charm against lightning and illness.

1

2

3

1. *Nerine sp.*
2. *Nerine angustifolia.*
3. *Nerine appendiculata.*

The *Brunsvigia* genus contains some of the most striking of our Drakensberg wild flowers. The plants consist of a large underground bulb and a stout, short stem, at the end of which is a large umbel of bright pink or rose-red flowers. Each separate flower with its six petals is on a short stalk, called a pedicel, and the petals curl back gracefully. When the flowers die they dry out, and the whole flower-head, stem and dried flowers, some 350 mm across, goes cart-wheeling over the veld in the wind, distributing the seeds over a wide area. The flowers are at their best during December and January.

The two main Drakensberg species are *Brunsvigia radulosa* and *Brunsvigia natalensis,* though it is possible that *B. undulata* and *B. grandiflora* also occur. The genus has two distinct types of leaves. In one, *B. radulosa,* the leaves are heart-shaped, rounded, and lie flat on the ground. The stem emerges from the side of the rosette of leaves, not the centre. In the other, *B. natalensis,* the leaves are long and strap-shaped, standing erect. These two types of leaves are clearly shown in the pictures opposite. The genus is essentially a South African one, being found nowhere else except for one species found in Central Africa.

Brunsvigia radulosa grows mainly amongst the rocks on the summit of the Little Berg, at a height of about 2 000 metres, though I have found it in grasslands at a height of 1 250 metres. *B. natalensis* only grows in grasslands on the lower slopes. A decoction of *Brunsvigia radulosa* is used by the Southern Sotho as a remedy for barrenness, while their diviners sometimes wash their divining dice in it in the belief that this imparts greater accuracy to them.

The name Brunsvigia honours the Duke of Brunswick (1735-1806). In the 18th Century many of the noble families in Europe used to collect foreign and exotic plants. Ryk Tulbagh, Governor of the Cape from 1751 to 1771, sent the Duke of Brunswick a bulb of *Brunsvigia.* It was named while flowering in the Duke's garden. *Natalensis* means of Natal, and *radulosa* comes from the Latin word *radix,* meaning a root. The Zulu name is um-Bhola, which means something rotten. The large bulb is usually full of water. This makes a layer of wet mud round the bulb, with a strong, rotten odour. Hence the Zulu name. Xhosas pound the roots and mix them with water to spray their huts with, so as to ward off evil spirits.

Although a most spectacular flower, Brunsvigias have never caught on with horticulturists, as the Agapanthus has. This is largely because of its slow rate of flowering. From the planting of the seed (which germinates freely) it takes some seven to eight years before the flowers appear.

One of the sights you will never forget in the Drakensberg is to find a colony of Brunsvigias, sparkling red against the dark rocks on the summit of the Little Berg in the bright sunshine of a summer's morning after a night of rain. When, in 1837, after months of trekking across the arid plains of the Free State, the Voortrekkers came to the edge of the Drakensberg, they were entranced by masses of large, rose-red, lily-like flowers growing amongst the summit rocks. These were almost certainly Brunsvigias. In their joy they called their camp *Bloemendal.*

1

2

3

4

5

The colour of *Brunsvigia natalensis* varies from a deep wine-red (No. 1 above) to pink (No. 2).

3. *Brunsvigia radulosa,* growing on the steep rock faces of the Little Berg.

4 & 5. *Brunsvigia radulosa*.

HYPOXIDACEAE

This is a small family, with less than a dozen genera and a little over 100 species. These genera used to be included in the Amaryllidaceae family, but some time ago they were placed in a family of their own. The family is distributed world-wide, except in Europe and northern Asia. The plants have six petals, an inferior ovary and tuberous rhizomes or corms. Only three genera occur in Natal, *Hypoxis, Rhodohypoxis* and *Curculigo.* I have never come across *Curculigo* in the Drakensberg, but the other two genera are common.

HYPOXIS

Hypoxidaceae

There are 21 species of *Hypoxis* in Natal, and at least seven of these occur in the Drakensberg. All have bright yellow flowers, with six petals, and most are known by the common name Geelsterretjies, or just Sterretjies. The flowers open only on bright, sunny days, and sometimes close in the afternoon, even though the sun is still shining. On a climb one morning I once came across a magnificent display of these flowers. I returned in the early afternoon with my camera, but even though the sun was still shining brightly, without a cloud in the sky, there was not a flower to be seen!

Many of the species are very similar to each other, and it is not easy to differentiate between them. *Hypoxis obtusa* is a robust plant, about 300 mm high. Its leaves are slender, bright green and glabrous. It flowers from October to December. The Zulus call it in-Komfe, and they pound the corms and boil them to make a decoction for stomach trouble. They also use the leaves, which are strong and fibrous, to make ropes and string.

Hypoxis rigidula is very similar to *H. obtusa,* but it can be distinguished from this latter species by its leaves, which are hairy, and the plant is more slender. (*H. obtusa* does have stellate hairs on the keel and margins of the leaves, but the lamina is glabrous). Like *H. obtusa* it is found in early spring, growing on the slopes of the Little Berg. The Southern Sotho name for it is tieane.

Hypoxis argentea is a dwarf Hypoxis, only a few centimetres high. Three other dwarf species, all occurring in the Drakensberg, are H. *parvula, H. filiformis* and *H. angustifolia.* The flowers of this latter species are yellow on the inner surfaces and green on the outer surfaces, and this differentiates it from the other dwarf species.

Hypoxis rooperi, the best known of the Natal species, has not been recorded so far from the Drakensberg, though it is widespread elsewhere in Natal. In Lesotho this species is used as a cure for headache. The rootstock is shaped into a hollow receptacle in which some blood from the forehead of the afflicted person is collected. The receptacle is then buried in the ground, and the headache is cured!

The name Hypoxis comes from two Greek words, *hypo,* meaning below, and *oxy,* pointed. This refers to the sharp points of the inferior petals.

1

2

3

1. *Hypoxis obtusa.*
2. *Hypoxis sp.,* probably *H. rooperi.*
3. *Hypoxis argentea.*
4. *Hypoxis obtusa.*

Rhodohypoxis is a genus of dainty little plants, well known to all who visit the high places in the Drakensberg. It is a small plant, only a few centimetres in height, with narrow, slender leaves. It is usually found growing in black, peaty soils, often shallow, and on moist rocky outcrops. Its favourite habitat is the summit plateau, at 3 000 metres, but it does grow as low as 1 500 metres.

The flowers are white, pink or rose-red, the colour sometimes being dependent on the acidity of the soil. They have a waxen sheen, and the plants are usually found growing together in small colonies, though white ones are often found growing singly in the Little Berg. The flower is six-petalled, and has six stamens.

To the uninitiated *Rhodohypoxis* has always had only one species, *R. baurii,* though botanists up to recently recognised three species and one variety. The most recent determination, however, gives us six species, one with three varieties. They are all endemic to the mountain regions of East Griqualand, Natal, Lesotho, the eastern Free State and the eastern Transvaal. They are as follows:

R. rubella, pink in colour, but occasionally white. The flowers are rather smaller than the other species. It is common in seasonally wet silt patches on the summit plateau.

R. baurii var. baurii, red in colour and usually found on dripping cliffs and rock sheets. It has not, as yet, been reported from the National Park and Cathedral areas.

R. baurii var. platypetala. This is the commonest species in the Drakensberg. It is usually white in colour, but is very occasionally pink.

R. baurii var. confecta. Flowers are red, pink or white, the white sometimes turning pink as the flower develops. The word confecta means, in Latin, completed or accomplished.

R. milloides. This is the old *R. palustris.* It is usually found in marshy places. The flowers are a deep, rich red, and the leaves are erect and bright green.

R. thodiana, a little more rare than the others. The flowers are pink in colour. It has been recorded from the upper reaches of the Bushman's River Pass.

R. deflexa. This is another very small flower, usually pink but occasionally white. It is found in marshes on the summit plateau.

R. incompta, another rare species. So far it has only been recorded from the summit of Sani Pass and in the Upper Polela area. Here the flowers are pink, and rather floppy in appearance. *Incompta,* in Latin, means inelegant, artless or rude.

I am indebted to Drs. O. M. Hilliard and Burtt whose revision of *Rhodohypoxis* was in the press when my own book went forward to the publishers, for this information.

As the name implies, the plant is a small relative of *Hypoxis.* It flowers from October to December, and has been cultivated extensively overseas, though not as yet in South Africa. According to Dr Jacot Guillarmod, overseas growers 30 years ago were prepared to pay up to £5 for a single tuber of the plant.

No-one who has ever visited the Sehlabathebe National Park in Lesotho, a few miles from Bushman's Nek, will ever forget the massed carpets of these very lovely little flowers, growing between the rock pools and amongst the crags of this remote and lonely Park. It is worth a trip to the Park if only to see this remarkable sight.

1

2

3

4

5

1. *Rhodohypoxis baurii var. platypetala.*
2. *Rhodohypoxis milloides.*
3. *Rhodohypoxis baurii var. platypetala.*
4. *Rhodohypoxis* hybridises easily. These are hybrids between *R. baurii* and *R. milloides.*
5. *Rhodohypoxis baurii var. baurii.*

VELLOZIACEAE

This is another new family which was originally included in the Amaryllidaceae. All the plants in this family have tough, central stems, draped firstly with numerous wiry roots, produced along the entire length of the stem, and then thickly swathed in the fibrous remains of the leaf bases. The leaves consist of two distinct parts, firstly a broad, sheathing base and then a long, grass-like blade. They are thickly crowded near the top, giving the plant a plumed appearance.

Distribution is from Brazil, through Tropical and Southern Africa, to Madagascar and South Arabia. In South Africa the family does not appear south of Botswana, Northern Free State and Natal. It is a small family, with only three genera. Only one, that of *Vellozia,* occurs in Natal.

VELLOZIA

Velloziaceae

Only three species of this family are found in Natal, and only two of these, *Vellozia viscosa* and *V. elegans,* are found in the Drakensberg. (Some authorities consider that the genus *Vellozia* is confined to the New World. If this is accepted, we would then have, in South Africa, *Talbotia elegans (Vellozia elegans)* and several species of *Xerophyta,* which would include *Vellozia viscosa).*

Vellozia viscosa is a purplish-magenta flower, usually found growing on flat, wet, slab-like stones, at the upper limits of the cave sandstone, though I have occasionally found it growing at lower altitudes, down to 1 250 metres. The plant grows to a height of about 250 mm, and the flowers are from 20 to 30 mm in diameter. Occasionally a white variety is found. Flowering time is from December to March.

Vellozia elegans (the old *Vellozia talbotii)* is smaller than *Vellozia viscosa,* about 100 mm high, and the diameter of the flowers is from 10 to 15 mm. It grows in shady places in the *Podocarpus* forests, usually on the sides of waterfalls, where it forms dense mats on the damp rock faces. Like all the other Vellozias, the plant has a tough, slim central shoot, enclosed in old leaves and wiry roots. The leaves have the usual plumed appearance. The flowers, which are dark blue with yellow centres, grow singly in the leaf axils. (It is this feature which made it necessary to place the Vellozias in a family of their own, for practically all *Amaryllidaceae* inflorescences are umbels).

The third *Vellozia* species, not found in the Drakensberg, occurs in Zululand, and also in the Transvaal, on the koppies around Johannesburg and Pretoria. This is a giant species, sometimes reaching a height of four metres. This is *Vellozia retinervis.*

The name Vellozia probably comes from the Latin word *vellus,* meaning something which covers, like wool or fleece, referring, no doubt, to the covering sheath of wiry roots. *Viscosa* comes from the Latin word for sticky, *viscosus,* and *elegans* is, of course the Latin word for choice, nice or showy.

1

2

3

1 & 2. *Vellozia viscosa.*
3. *Vellozia elegans.*

IRIDACEAE

This is a large family, with something like 75 genera (43 coming from South Africa) and 860 species. It was named after Iris, Goddess of the Rainbow, because of the many beautiful colours of the flowers in this family. Distribution is throughout tropical and temperate regions, mainly Africa and America. Apart from the 'petals', all the flower parts, including the stamens are in threes (contrasting with the Liliaceae and the Amaryllidaceae, which are in sixes). The ovary is inferior, i.e. the perianth segments, stamens etc. are inserted above the ovary. The underground system is usually a corm. (Corms are like bulbs in shape, but have a solid centre and are more compressed. The new corms arise on top of the old ones). South Africa is particularly rich in Iridaceae flowers. All our Freesias, Watsonias, Tritonias, Gladioli, Ixiae, Babianas and Moraeas are members of this family.

SCHIZOSTYLIS

Iridaceae

The scarlet flowers of *Schizostylis coccinea,* nodding their heads over the clear waters of a mountain stream, always excite the admiration of visitors to the Drakensberg in the late summer months. These flowers may be found growing on the banks of streams in the Little Berg during the months of January and February. The six star-like pointed petals glisten in the sunlight with a satin-like sheen, and the leaves, dark-green, long and pointed, make a fitting foil to this lovely flower. The flowers are borne at the tips of slender stems from 0,4 to 0,5 m in length, and the flowers themselves are about 40 mm in diameter.

Exploring one of the remoter valleys of the Little Berg one Easter, I came across a pink variety of the flower, which, I believe, is somewhat rare. I have actually seen it only twice in the Drakensberg, once on this occasion, in the Setene valley, and once several years later, beyond Castle Rocks in the Royal National Park area. It used to be called *Schizostylis pauciflorus,* but the name has now been dropped and the pink variety is now included in *Schizostylis coccinea.* This is the only species in the genus, which is confined to South Africa.

The name comes from the Greek word *schizein,* to split, (referring to the split style), and *coccinea,* the Latin word for scarlet. Common names are Crimson Flag, Kaffir Lily, Kafferlelie, and River Lily.

The flower has been a favourite overseas, where it is widely cultivated. According to Conrad Lighton, it has been a tremendous success in the Laird of Inverness' waterside garden on the north-west coast of Scotland (where, incidentally, he has also grown with success our *Phygelius capensis).* "One sees its masses of dazzling scarlet on my terrace from a boat sailing about in the bay", says the Laird in his Diary, "A Hundred Years in the Highlands", quoted by Conrad Lighton in his book, "Cape Floral Kingdom".

Schizostylis coccinea

As one climbs out of the lip of the Mnweni Pass in the Drakensberg, in the height of midsummer, it is not unusual to find great masses of yellow colour against the dark grey of the basalt. These are Moraeas, and the specific name for this particular species is *Moraea spathulata*. Actually, the Moraeas are well represented in the Drakensberg, no less than 13 species having already been recorded there. *Moraea spathulata* is probably the most widespread. Its common name is the Yellow Iris, and it is very similar in appearance to the cultivated Bearded Iris. The plant is from 0,6 m to a metre tall, with narrow, ribbed leaves and large yellow iris-like flowers at the tip of a stout stem. It is found growing all along the summit of the Drakensberg, especially in damp places, and I have seen acres and acres of them there, making a glorious splash of colour.

The Zulu name for the plant is i-Hlamvu elincane. For a woman who fails to conceive the corm is ground up and mixed with two or three seed mealies. From these, three small cakes are made. She eats two, her husband eats one, and, hey presto! – it isn't long before the patter of little feet is heard! But there is also a more sinister side to the plant. It is poisonous, especially to stock (the common name is Tulp), but the exact nature of the poison has so far eluded scientists in spite of intensive work on the plant for many years. Tests at Onderstepoort have proved that half a kilogram of Tulp, green or dried, can kill an ox in 36 hours. Strangely enough, however, the plants are eaten in large quantities by rodent moles without any ill effects. Porcupines, also, are said to be immune to it. The plant is found throughout the Eastern Province, the Transkei, Natal, Lesotho and the Transvaal.

1

2

3

1. A field of *Moraea spathulata* in the Mont-aux-Sources area.
2. *Moraea spathulata*.
3. *Moraea spathulata*, at the head of the Mnweni Pass.

Moraea stricta (up to recently this was called *Moraea trita)* is a somewhat smaller flower, only about 0,2 m high, and with flowers about 25 mm across. The flowers are mauve in colour, with a yellow and white spot towards the base of three of the six petals. The stems are thin and wiry.

Moraea huttonii grows along stream banks at about the 1 500 to 1 800 metre levels. It flowers from July to December, the flowers are yellow in colour and are slightly smaller than *Moraea spathulata.* It used to be called *Moraea rivularis.*

Moraea alticola grows only on the summit of the Drakensberg, as far as I know. This species is another yellow one, and is very similar to *M. spathulata.* It is, however, larger and more robust. It has been recorded from the Mont-aux-Sources area, and whole fields of them may be seen on the summit of the Sani Pass.

A strange coincidence occurred in the naming of this genus. A botanist gave it the name *Morea* in honour of Robert Moore, squire of Shrewsbury and a plant collector in the 18th century. It was subsequently discovered that Linnaeus had already named the genus *Moraea* after his wife Sara Elizabeth Moraea, daughter of the town physician of Falun, Dr Johan Moraeus. The pronunciation was the same, but the spelling different. In accordance with the International Rules of Nomenclature, the original spelling *Moraea* had to be reverted to.

1. *Moraea alticola.*
2. *Moraea alticola.*
3. *Moraea stricta.*
4. *Moraea huttonii.*
5. *Moraea spathulata.*

1

2

3

4

5

The flowers of the genus *Dierama* are amongst the most beautiful of our Drakensberg wild flowers. Dieramas all have thin, rigid leaves resembling a grass, and a slender, wiry grass-like arching stem, with a series of lax spikes near the tip, each bearing several pendulous bell-like flowers. The Greek word *dierama* means bell-like. They bloom from October throughout the summer, and vary in colour from white, through mauve and pink to purple, magenta and wine-red. The buds are covered with papery bracts, and the root-stock is a corm covered with dry fibres. Common names are Wand-flower, Harebell, Fairy Bell, Grassy Bell, Grasklokkie, Wedding Bell and Flowering Grass.

For some time there has been considerable confusion amongst botanists as to the specific names of this genus. Up to the early 1920's only two species and one sub-species were recognised, *Dierama pendula, D. pulcherrimum* and *D. pendula var. pumila.* Up to quite recently many botanists were of the opinion that this represented the sum total of species, but they recognised many varieties. At the latest count (1976) twenty species were recognised, but it would be difficult to say how many of these occur in the Drakensberg.

One, certainly, and a very interesting one, is *Dierama igneum.* This is a dwarf species growing on the summit, and it is an exquisite little flower. It is only about 180 mm high (most Dieramas grow to a height of about 0,5 m), and it has two colour varieties, a rich, deep red and a dark mauve. Its small size has interested botanists. At high altitudes there is an increase in the intensity of ultra-violet rays. These rays inactivate the growth-promoting hormones, and it is thought that this results in the dwarfing of many species of high-altitude plants.

Conrad Lighton, in his book "Cape Floral Kingdom", describes how, many years ago, South Africa sent *Dierama* seeds over to Ireland. They were planted in County Down, where the Mountains of Mourne sweep down to the sea. Here, in the rich, moist soil, strangers in a strange land, they found a natural home, and if, to-day, you visit County Down you will find whole fields of them, nodding their dew-laden heads in the breeze. Spikes of up to 2,75 metres have been measured.

In the Drakensberg I know of at least four species, *Dierama igneum, D. robustum, D. medium* and *D. trichorhizum.* They are often found in the upper reaches of the passes leading to the summit, and on the summit itself the sight of *Dierama igneum,* nodding its rose-red bells over a rippling mountain stream, is enough to give a lilt to your heart that you will long remember.

Dieramas do not appear to have many medicinal uses amongst Africans. A decoction made from the corms of *Dierama pendula* is used by the Southern Sotho as a strong purgative enema, while the Zulus, who call it i-Thembu, use *Dierama robustum* for its fibre in the making of cords and thread.

1

2

3

4

5

1. *Dierama sp.*
2. *Dierama igneum.*
3. *Dierama igneum*
4. *Dierama robustum.*
5. *Dierama igneum.*

HESPERANTHA

Iridaceae

In your wanderings in the forests of the Drakensberg you will often find, growing on the sides of waterfalls and on wet rock faces, a dainty, scarlet flower with six petals and long, thin leaves. This delightful little flower is almost certain to be *Hesperantha longituba*. It flowers in March and April, sometimes only in the afternoon. (The name Hesperantha means evening flower and the Afrikaans common name is Aandblom or Avond-blommetjie). Also, in the evening, the flowers are strongly scented, at a time when night-flying insects, so necessary for their pollination, are abroad. The Zulu name is isidwe.

Another species, which you will also find in early autumn, but in the grasslands, this time, of the Little Berg, is *Hesperantha baurii*. This is a particularly lovely little flower. It is a deep, rosy pink in colour, the flowers growing on a slender spike, and the whole plant is about 150 mm high. An infusion made from the pounded corms of this plant is used by the Zulus as a cure for chest troubles.

There are about 50 species of this genus altogether, nearly all of which are found in South Africa. Eleven occur in Natal. Most of them are pink or red in colour, but there are a few white ones.

DIETES

Iridaceae

Dietes iridioides (the new name for the old *Dietes vegeta)* is a flower which you will not often see in the Drakensberg, not because of its rarity – it is fairly widespread in the forests of the Drakensberg – but because it flowers rarely and the flowers last for only one day. There are six species to the genus altogether, five in Southern Africa and one in Australia. In South Africa it has been recorded from the eastern Cape, the Transkei, Natal and the Transvaal.

As far as I am aware, only *Dietes iridioides* occurs in the Drakensberg. It is essentially a shade-loving plant, and is usually found in the depths of the *Podocarpus* forests. The flowers are very similar to those of the Moraeas (Iris) and at one time were actually classified under *Moraea*. Their root system, however, is not a corm, as in *Moraea,* but a sturdy, creeping rhizome, and it was thought better to create a separate genus for these plants.

The flowers have six petals and are pure white, faintly tinged with blue, with a yellow spot on the three outer petals. They measure about 40 mm in diameter, and the plant grows to a height of about 0,6 m. Flowering time is mid-summer. The leaves, long, sword-shaped and evergreen, are in the shape of a fan, and the forest floor is often carpeted with them. *Dietes iridioides* has been known in cultivation for at least 200 years.

The name Dietes comes from two Greek words, *di* meaning twice, and *etes* which means an associate, the word apparently referring to the petals. The Zulus make an infusion from the inner part of the corm and use it in the treatment of dysentery. Their name for the plant is i-Ndawo yehlathi.

1

2

3

4

1. *Hesperantha baurii*
2. *Hesperantha longituba*
3 & 4. *Dietes iridioides.*

The genus *Aristea,* with 49 species, is essentially an African genus, but six species are found in Madagascar. The rest are all African, with eight coming from Natal and probably four being found in the Drakensberg. Aristeas are almost invariably a rich, sparkling blue in colour, but there are one or two fairly rare white species.

There is some doubt as to the exact derivation of the word Aristea. Some authorities believe that the word comes from the Latin word *arista,* meaning the top, or beard, of an ear of corn, and refers to the appearance of the pointed leaves. A more likely derivation, however, is that it comes from the Greek word *aristos* which means best, referring to the attraction of the flower. (Compare our word aristocracy, which is also derived from the Greek *aristos,* and which the Concise Oxford Dictionary defines as government by the best citizens). In Greek, the word *aristeia* indicated excellence, and was the name of the prize given to the best and the bravest soldier in the Second Persian War.

The flowers of most species open only in the morning, closing in the afternoon, and if picked they twist up into a spiral and turn black. Flowering time is late October to January, and in South Africa they are found in Natal and the Transvaal, and along the coast into the Cape up to Clanwilliam. Being a member of the Iridaceae family the flowers have six perianth lobes. Some are arranged in a spike and some appear singly at the tip of the stalk.

I know of four Drakensberg species, *Aristea angolensis, A. cognata, A. woodii* and *A. ecklonii.* *Aristea cognata* is a small plant with a single flower borne at the tip of a slender stem about 100 mm high. It may be found in the grasslands of the Little Berg, and it flowers in the late summer. Common names for it are Small Blue Iris and Miniature Iris. The Zulu name is icebetwane. The Zulus make an infusion from the leaves and this is used as a hot fomentation for sprains.

Aristea ecklonii has long, thin leaves, about 350 mm in length, usually eight in number and arranged in a fan. The flowering stem is branched and is up to 450 mm in height, bearing flowers on short 5 mm stems. The flowers are about 20 mm across, and they close in the afternoon. It is often found growing in damp places on the edges of forests, and the flowers appear in spring and early summer. Common names are Blouknol, Blue Day Lily and Koringblommetjie.

Aristea woodii has flowers which are a rather paler blue than the other Drakensberg species. The flower stem grows up to a height of 0,45 m, and the flowers, up to 20 mm across, are in the form of a spike. Like *A. ecklonii,* it grows in semi-shade on the forest edges, and flowers from December to March. The Zulu name is umluzi omncane, and the Zulus make an infusion from the roots and leaves. This is used by young men for bathing the body when courting.

With their sparkling blue colour, shining in the sun, these flowers will always delight the heart of the mountaineer as well as the hiker who roams the sheltered valleys of the Little Berg.

Aristea woodii

Tritonia is essentially a Cape genus (the famous Mossel Bay Kalkoentjies are Tritonias), flourishing especially in the south-western Cape. In Natal we have seven species to our credit.

The flowers, growing in the form of a long spike, are in a sort of half-twist. They have a short tube, opening into six 'petals'. The spike is one-sided, and usually curls over to one side. Leaves are long and narrow, and arranged in the form of a fan. The rootstock is a corm.

Mr Ker, the author of this genus, tells us how he came to hit on the name Tritonia for it. He gave it the name because the stamens and the style seem to take up different positions in the flower, and are as variable as a Triton on a weather-vane. (Triton was a minor Greek sea-god who used to blow through a shell to calm or rouse the sea. Weather-vanes often carried the figure of the God on them).

Only two species occur in the Drakensberg. *Tritonia lineata* is a creamy-white to pale yellow species, found on the lower reaches of the Little Berg and flowering in September and October. The flowers are easily recognised as they are veined with thin, dark pencil-lines (hence the specific name) and the anthers have a black line down the sides. The slender stems are from 0,4 to 0,5 m long, and the flowers are about 20 mm in diameter. The rootstock is a small corm, and the leaves, stiff and slender, grow to a length of 150 mm.

Common names are Bergkatjetjie and Pencil Tritonia. The Zulu name is isi-Dwa esimpofu. Roots are burnt and ground to a powder by the Zulus and this is used to heal an infant's navel if it is not healing naturally.

Tritonia rubro-lucens (at one time it was called *Tritonia rosea)* is a rose-red Tritonia, with stems about 0,3 m long. It is usually found growing along stream banks in the Little Berg, and flowers in January. The specific name comes from two Latin words, *rubro* meaning red, and *lucens* meaning bright or luminous. The Zulu name is isi-Dwa esibomvu. An infusion is made from the crushed corms by the Zulus, and this is used as a remedy for newly-born infants suffering from stomach trouble.

Tritonias are very lovely flowers. Not only are they cultivated extensively in our gardens, but, crossed with *Crocosmia,* they have given rise to our garden Montbretias. There are about 40 species altogether, all coming from Southern Africa.

1

2

1. *Tritonia lineata.*
2. *Tritonia rubro-lucens.*

There is considerable uncertainty amongst experts on the matter of our Drakensberg Watsonias, and the genus is definitely in need of revision. At one time it was held that only two species occurred in these mountain regions, *Watsonia densiflora* and *W. meriana*, but J. W. Ross, in his "Flora of Natal" (a standard work) (1972) listed three, *W. densiflora, W. meriana* and *W. lepida*. Dr D. J. B. Killick (1963) collected *W. lepida* and *W. socium* in the Cathedral area. The latest determination would indicate that, as originally assumed, only *W. densiflora* and *W. meriana* occurred in the Drakensberg, but that they varied considerably in growth form and in colour, and that *W. lepida* is a synonym of *W. densiflora* and *W. socium* a synonym of *W. meriana*.

Watsonias are tall, handsome plants, sometimes up to a metre in height, with stout stems and narrow, sword-like leaves. Colours range from white to coral pink, red and orange.

Watsonia densiflora, perhaps the commonest of all Watsonias, is a bright, pinkish-mauve in colour, and particularly common in the Drakensberg. It is easily recognised both from its colour and from the arrangement of the individual flowers on the spike. These flowers are borne in dense clusters, arranged in two opposite ranks up the stem. This gives the spike a flattened appearance. The flowers emerge from a core of long, brown tightly-packed imbricate bracts. These tightly-packed bracts are a distinguishing feature of the species.

The plants usually form small, compact groups, as in the third picture on the next page, which was taken in the foothills of the Drakensberg, in the Giant's Castle area, but in the Cathkin area the plants are quite different. Here they grow in a single spike, with a minimum of leaves, and they flower in early September and October, while the bushy type flowers in late December and January, and has a proliferation of leaves.

W. meriana is also well-known in the Drakensberg, where it grows at a slightly higher altitude than does *W. densiflora* (about 1 400 m) and usually amongst rocks and boulders, again in contrast with *W. densiflora,* which grows in grasslands. The prevailing colour is a bright orange (in the Cape the colour is usually coral or salmon-red) and the flowers are large and funnel-shaped, with flaring lobes. They appear much more sparsely on the spike than in the case of *W. densiflora*.

In the Drakensberg we have one very rare colour variety of *W. meriana,* a lovely carmine to deep crimson, a most beautiful flower. In all my years in the Drakensberg I have only seen it once, in a sheltered valley of the Little Berg in the Cathkin area. (The flowers are arranged rather differently on the spike to the orange *W. meriana,* and to my uninitiated eyes the plant looks like a different species, but it was identified definitely as *W. meriana).*

Watsonias are very similar in appearance to Gladioli, and species of the two genera are sometimes mistaken for one another. The main difference between the two genera is that in *Watsonia* the three branches of the stigma are each divided further into two smaller branches, while these branches are not divided in *Gladiolus*.

Common names for *W. meriana* are Lakpypie, Knolpypie, Rooiknolpypie and Suurknol. The Zulu

1. *Watsonia meriana,* the rare crimson variety.
2. *Watsonia meriana,* the common orange variety.
3. *Watsonia meriana,* growing amongst the rocks of the Little Berg.

1

2

3

name for *W. densiflora* is intshumo, and for *W. meriana* isidwe esincane. Mairn Hulme, in her "Wild Flowers of Natal" tells how two or three Zulus would sit together smoking dagga through an igudo, a smoking horn, using the stem of a Watsonia for this purpose, which they would hand to each in turn. Each man would take a deep draw, filling his mouth with the smoke. He would then take up the flower stem and blow spittle through it on to the floor of the hut, making patterns with the bubbles full of smoke. The game was to try and make the most intricate pattern, and this became quite complicated, the game continuing until one or the other proved his superiority at the art. Tshuma, in Zulu, means to spit through a reed while smoking, hence the Zulu name for the plant.

There are about 70 species of *Watsonia* altogether, and the genus is found growing wild only in South Africa and Madagascar, usually in the mountainous areas. The genus was named 200 years ago by Philip Miller, Curator of the Chelsea Botanical Gardens, after Dr William Watson, a London apothecary. Watsonias were introduced to England and the Continent many years ago from the Cape. They were growing in the Chelsea Gardens as far back as 1759.

There is one intriguing flower worth mentioning here, for there is a mystery attaching to it. It is extremely rare (I have seen it only twice in the Drakensberg but other mountaineers have occasionally reported having seen it, rarely). It is a brilliant orange flower with a spike rather similar to that of *Crocosmia paniculata*. The individual flowers, too, are rather like those of *C. paniculata,* long and tubular. It stands about 0,3 m in height, and grows on the slopes of the Main Berg at altitudes of between 2 200 and 3 000 metres. It flowers in January. Two Herbaria, to whom I have sent specimens, have been unable to identify it, and it may be a new genus or species. How such a striking and brilliant flower could have escaped the attention of collectors in the past is difficult to understand, and is part of the mystery. The specimen opposite was photographed at the head of the Mnweni Pass, at an altitude of 2 897 metres, in January 1977.

1

2

3

4

1. *Watsonia densiflora*.
2. *Watsonia meriana*.
3. The mystery flower.
4. *Watsonia densiflora*.

CROCOSMIA

Iridaceae

One of the showiest flowers in the Drakensberg in the late summer is *Crocosmia aurea,* a lovely name for a strikingly beautiful flower. The generic name comes from the Greek word *crocus,* meaning saffron, and *osme,* meaning scent, while the specific name comes from the Latin word for golden, *aureus*. You will find this lovely flower in the shade of the *Podocarpus* forests, especially along the banks of streams, in late summer. The plant is characterised by large, bright, golden-orange flowers in a full spike at the head of a tall stalk, from 0,3 to 0,5 m high. The flowers hang down slightly, with particularly graceful stamens. The long, tapering leaves, with a pronounced mid-rib are arranged in a sort of fan, coming up on either side of the scape. Its common name is Falling Star.

In shape the flower is very similar to that of a Tritonia, and in fact the plant was originally classified under this genus. This is no longer accepted, and it is now classified as *Crocosmia*. Up to recently there were only two species in Natal, *Crocosmia aurea* and *C. pottsii,* both of which were found in the Drakensberg. *C. pottsii* is somewhat smaller than *C. aurea,* and has a smaller, darker flower. In addition to its usual forest habitat *C. pottsii* is sometimes found along road-sides. In this species the perianth segments are straight and not reflexed as in *C. aurea*.

Recently, however, the well-known *Curtonus paniculatus* was re-classified as *Crocosmia paniculata,* giving three species to the genus in Natal. *Crocosmia paniculata* is a tall perennial, growing in clumps, with sword-shaped, pleated leaves tapering at both ends (its common name is actually Pleated Leaves), with branched inflorescence. Each branch ends in a many-flowered spike, with a sort of zig-zag axis. The flowers are tubular, with protruding stamens. It is often found growing immediately below the sandstone cliffs of the Little Berg. There are two colour varieties, a brownish-gold and a rich red. It flowers during February and March, and its Zulu name is um-Lunge. *Paniculata* comes from the Latin word *panniculus,* meaning a small piece of cloth.

Crocosmia paniculata occurs in the Transvaal, Swaziland, Lesotho and Natal, while *Crocosmia aurea* is found from the Cape, along the coast and into Natal, the eastern Transvaal and on into tropical East Africa.

Crocosmia paniculata is a Sotho remedy for diarrhoea.

GLADIOLUS

Iridaceae

A rather charming Greek legend tells the story of two brothers who fell in love with the same girl. Their love for each other turned to hatred. They fought, and each was mortally wounded. Realising too late their folly and that death was upon them, they drove their swords into the ground in peace, and immediately a flower emerged, with sword-like leaves – the flower of Peace, the Gladiolus. The name comes from the Latin word for a sword, *gladius*.

The genus *Gladiolus* is one of the largest in the family of the Iridaceae. It is widespread, occurring in Europe, the Mediterranean region, western and central Asia, and of course South Africa. Two species are endemic to Madagascar, and this may indicate that the *Gladiolus* had established itself well before the break-up of ancient Gondwanaland, 150 million years ago. There are probably about 150 species altogether, and of these about 105 come from South Africa. Natal contributes 19 to this total, and 17 of these are known to occur in the Drakensberg. It is impossible to deal with them all in a book of this nature, and I have selected six of the better-known species for inclusion here.

Possibly the most brilliant of all Gladioli is one that is endemic to the Drakensberg, *Gladiolus cruentus*. It is found nowhere in the world except along a narrow strip on the summit of the Drakensberg, from Mont-aux-Sources to round about Giant's Castle. The flowers are bright scarlet, and the petals have a mottled throat. The word *cruentus,* in Latin, means blood-stained. The stem is usually curved. It grows high up on steep cliff faces, in inaccessible crevices, and because of this is commonly known as the Suicide Gladiolus. It is, indeed, a magnificent flower.

Gladiolus cruentus must not be confused with another red *Gladiolus,* which is also found on cliff faces in the Drakensberg, *Gladiolus saundersii*. This species is more vermilion in colour than the last, with the three lower lobes tinged with white or cream. The two species also differ in the shape and set of the flowers. In *G. cruentus* the tube is either straight or slightly curved. In *G. saundersii* the tube is strongly curved. *G. saundersii* is the more widely-spread of the two, being found in the eastern Cape, Lesotho and the Natal midlands, as well as in the Drakensberg.

Gladiolus crassifolius is also widely distributed,

1

3

4

2

5

1. *Crocosmia aurea*
2. *Crocosmia paniculata.*
3. *Crocosmia paniculata.*
4. Ditto: the orange variety.
5. Ditto: the scarlet variety.

being found all the way from the Eastern Province, through Natal, Lesotho, the eastern Orange Free State, to the Transvaal, and possibly to Rhodesia and Moçambique. In the Drakensberg it grows at about the 1 250 metre level, and flowers during February and March. The flower spike has a number of small, closely-compacted pinkish to mauve flowers. It is easily identified by reason of the fact that the stem usually curves out at right angles from the stiffly erect leaves. Characteristically, also, the leaves have yellow, raised ribs. In Lesotho it is called khahla e nyenyane, meaning the small thing which pleases. A preparation from it is used as an enema and as a cure for headaches and lumbago. As a species it is somewhat smaller and daintier than the others. H. M. L. Bolus, however, of the Bolus Herbarium, gave it the name Thick-leaved Gladiolus. The word *crassifolius* comes from the Latin word, *crassus,* meaning thick.

Gladiolus longicollis var. longicollis is a pale yellow *Gladiolus,* with brown streaks down the petals. Very typical are the tapered apices of the petal lobes. Once common in the Eastern Province of the Cape, it is now almost extinct there. This may be due to the fact that the corms are often eaten by partridges and small animals. To-day it is found only on the summit of the Drakensberg, occasionally in the Little Berg, rarely in the Midlands of Natal, and in parts of Lesotho. Its common name is the Honey Flower, and the Zulu name is umbejo.

Gladiolus natalensis is probably one of the best-known of our *Gladioli,* and it is widely cultivated. It used to be called *Gladiolus psittacinus,* the name coming from the Latin word for a parrot, *psittacus,* and was often known as the Parrot Gladiolus, as the flower looked like the beak of a parrot. The name has now been changed to *G. natalensis.* The epithet *psittacinus* had been given as far back as 1830, and had persisted until quite recently, but the species had already been named *natalensis* by Reinwardt, and the earlier name had to be restored. It is probable that corms of the plant were first brought to the Cape, and thence to Europe, from Natal after Farewell's landing there in 1823.

It is a robust plant, growing up to 0,6 m in height, though sometimes the flowering spike has been known to reach a length of over a metre. Because of its wide distribution (it is found in all four Provinces of the Republic as well as in Swaziland, Moçambique, S.W. Africa, Botswana, Tropical Africa, Ethiopia and western Arabia), it differs considerably, especially in colour, from one area to another. To add to the confusion,

---

1. *Gladiolus longicollis var. longicollis.*
2. *Gladiolus daleni.*
3. *Gladiolus cruentus.*

1

2

3

many other species of *Gladioli,* such as *dracocephalus, primulinus, quartinianus, tysonii,* in addition to *psittacinus* have now been merged into *G. natalensis,* and extensive hybridisation has taken place. Its colour is usually a rich orange, but it can grade into cream, yellow and red.

Gladiolus daleni, named after Dr Cornelius Dalen, Director of the Rotterdam Botanic Gardens round about 1827, is a Gladiolus which is found in the valleys of the Little Berg, often along stream banks. The flowers are a deep orange, flecked with red, and they appear during January and February.

ORCHIDACEAE

The Orchidaceae family is one of the largest of the Monocotyledons. There are something like 725 genera altogether, and over 20 000 species. In South Africa we have 50 genera and 550 species. It is distributed world-wide, except in the Arctic and Antarctic regions. In the Tropics the plants are mostly epiphytes (i.e. plants growing on other plants, but not parasitically). Most of the South African species are terrestrial, or ground, orchids. Many of the epiphytes have been hybridized and cultivated as prized green-house flowers.

The family is considered to be the most highly developed of all the Monocotyledons, and the flower structure is extremely complex. It is designed specifically to assist pollination by insects. The flowers are irregular, and basically consist of six perianth segments in two whorls of three segments each, three outer and three inner. One of the inner segments becomes larger than the rest, and is usually different in colour, texture and shape. This is called the lip, or labellum. The labellum is continued at the back of the flower into a typical spur, which, in the case of *Satyrium,* is split into two spurs. The six stamens and the style are united into what is called the column.

The word Orchid comes from the Greek word *orchis,* meaning a testicle, referring to the shape of the tuber.

---

1. *Gladiolus crassifolius*
2. Ditto. Note the colour variation.
3. *Gladiolus natalensis*
4. *Gladiolus cruentis*

1

2

3

4

SCHIZOCHILUS

Orchidaceae

Schizochilus flexuosus is a small white ground orchid, growing to a height of about 150 mm on the lower slopes of the Little Berg. The labellum (the lowest petal, facing towards the front) is yellow, and the flower is about 12 mm in diameter.

Schizochilus is an African genus of about 20 species, with two coming from Tropical Africa and the rest from South Africa. I know of three that occur in the Drakensberg, the other two being *S. angustifolius* and *S. zeyheri*.

The name Schizochilus comes from two Greek words, *schizein,* to split or cleave, and *cheilos,* a lip. *Angustifolius* means narrow-leaved, while the word zeyheri honours the well-known South African pioneer botanist, Carl Zeyher, who collected in the Cape round about 1822.

This lovely white flower was photographed late one afternoon in the Little Berg after a summer's day spent in the high hills, on my way home.

SATYRIUM

Orchidaceae

The genus *Satyrium* is another ground orchid. It is easily distinguished from the other Orchids by the two spurs of the hood-shaped lip at the back of the flower. All other genera in this family have only one spur. In general the plants are fairly small, and the inflorescence is in the form of a single spike. They are usually found growing in grasslands on the lower slopes of the Little Berg, though some species are found on the summit plateau of the Main Berg.

Satyrium is a large genus, with something like 137 species, mostly from Africa. Natal can lay claim to 11 of these, while I know of at least six in the Drakensberg.

Satyrium longicauda is one of the best known of these ground orchids. The plants are usually found growing in groups, often in damp places. Stems are about 0,3 m high, and the small flowers are closely compacted around this central stem. There are two colour varieties, a cream, and a white one with a red patch down the centre of the petals. The rootstock is a large unbranched tuber, and the leaves, growing near the base, are usually two to three in number and are pale green in colour. The flowers are wax-like and sweetly scented. They usually appear during January. The picture opposite was taken on the summit plateau of the Main Berg in the Mnweni area, in January. This locality is rather unusual, for the plant occurs mainly from the coast and up into the Drakensberg to a height of about 2 500 metres. The two spurs which differentiate this genus from the rest of the Orchidaceae are particularly prominent in this species. They are longer than usual, reaching a length of 20 mm. *Longicauda,* coming from two Latin words, *longus* and *cauda,* means long-tailed.

The genus was named after the Greek Satyr, a class of woodland deities, half man but with the legs of a goat, some of whom were supposed to have two horns.

---

1. *Schizochilus flexuosus*
2. *Satyrium ocellatum*
3. The pink spike in the centre is *Satyrium neglectum.* On either side, *Habenaria dives,* another of the Orchid family.
4. *Satyrium longicauda*

1

2

3

4

The genus *Disa* is mainly a coast genus, but in Natal it is found much more commonly inland, and in our mountain regions it is found frequently. At least a dozen species have been identified in the Drakensberg.

The flowers are usually small, and, like *Satyrium,* are closely packed on a central spike. The individual flowers are highly specialised. They are arranged with an upper apical sepal which is hooded and has a backward pouch or spur (not split as in *Satyrium)* and two lower lateral sepals spreading out sideways. There is an inner whorl of narrower petals of which the lowermost forms the lip.

Easily the best-known species of the genus is the strikingly beautiful *Disa uniflora* of the Cape, the Pride of Table Mountain. This brilliant scarlet flower has been chosen as the emblem of the Mountain Club of South Africa. It is found only in the mountainous regions of the south-west Cape.

Disa stachyoides, illustrated opposite, is a small Disa, about 0,25 m high, growing in grasslands in the Drakensberg at an altitude of about 1 800 metres. The flowers are red in colour, and they appear in January and February. The leaves are long and pointed. The Zulu name is i-Hlamvu elimpofu yasenkangala. An infusion is made from the entire plant by the Zulus and used to ward off evil and storms.

Space does not permit of our listing all the species of *Disa* that occur in the Drakensberg, but we might notice one other that is fairly common, *Disa rhodantha.* This is one of the most attractive of the Drakensberg Disas. It is usually found in marshes on the summit, at 3 000 metres. The flowers are mainly rose-pink, but the lip is light yellow to greenish in colour.

Disa was named by Petrus Jonas Bergius, a medical doctor and scientist from Stockholm, in 1764, but there is some doubt as to the origin of the name. It may come from the Greek word *dis,* meaning rich. Conrad Lighton, in his "Cape Floral Kingdom" quotes Dr L. Bolus as suggesting that Perdita's well-known flower speech in Shakespeare's "The Winter's Tale" might hold a clue to the meaning of Disa. "O Proserpina", says Perdita, "for the flowers now that, frighted, thou let'st fall from Dis's wagon" – but surely a rather unfortunate derivation, for Dis (Proserpina's husband) was Pluto, the god of Hades and of death! A more likely derivation might be that it is an allusion to the mythical Queen Disa of Sweden who was commanded to come before the King of the Sveas neither naked nor clothed, and who accordingly came wrapped in a fishing-net. (To-day she would no doubt have appeared in a bikini!) This is probably the correct derivation. The upper lip of the flower has a somewhat netted appearance, and Bergius was a Swede.

Disas are essentially mountain flowers. After quoting Dr Rudolf Marloth, the great South African botanist, as remarking of the Disas that their beauty "can only be fully appreciated when seeing the blooms on a cliff, just touched by the morning sun, or on the banks of a murmuring brook whose clear water reflects their glory", Conrad Lighton says in the book to which we have already referred "The promise of the sight of Disas in their natural setting has for centuries lured man to the heights. Life and limb have been risked for the fleeting ecstacy, which afterwards, paradoxically, endures as a lyrical memory".

1

2

3

1. An undescribed species of *Disa,* found on the slopes of Amphlett in the Cathkin area.
2. *Disa stachyoides.*
3. A field of Disas in the foothills of the Little Berg near Giant's Castle.

PROTEACEAE

The family of the Proteaceae is a large one. It contains something like 60 genera and 1 400 species. It is confined almost entirely to the southern hemisphere, and is largely concentrated in Southern Africa, particularly in the Cape Peninsula and surrounding areas. Practically all plants in this family are shrubs or trees, and it includes the well-known Leucospermums (Pincushions) and Leucodendrons (the well-known Silver Trees of the Cape). The genus *Protea* is the largest and most important South African genus in the family.

PROTEA

Proteaceae

The Protea, of course, is South Africa's national flower, a happy choice, for it is amongst the most striking of all our wild flowers. Actually it was the first South African indigenous flower to be known and illustrated overseas. This was *Protea neriifolia,* illustrated, without a name, in Clusius' "Exoticorum libri decem" in 1605, nearly fifty years before van Riebeeck landed in South Africa. The plant itself had been taken over in 1597. It is generally held that the genus was named by Linnaeus in 1735, after the Greek god Proteus, who was able to change his shape into many different forms, this referring to the fact that Proteas occur in such a bewildering variety of shapes and forms. But Professor W. P. U. Jackson, in his "Wild Flowers of Table Mountain", points out that Linnaeus had seen only a few Proteas when he named the genus, and could not have known of its variability. Professor Jackson thinks that Linnaeus selected the name haphazardly, but admits that it was a very lucky choice.

The Protea 'flower' actually consists of a mass of tiny flowerets, in the form of a rounded saucer, surrounded by coloured bracts. It is these bracts which make the Protea such a magnificent flower. Each tiny floweret has a perianth of four segments and four stamens.

In South Africa the genus is conveniently divided into two main groups, the winter rainfall Proteas of the Cape and the summer rainfall Proteas of Natal and the Highveld. It is in the Cape, of course, where they flower in their full glory. In the Drakensberg we have five species, and, fortunately for the beginner, these are easily identified.

Protea roupelliae is the commonest of all Drakensberg Proteas. It is a fairly large tree, growing to a height of four to five metres. The young leaves are soft and covered with woolly down, which gives the leaf a bluey-grey, silvery sheen. As the leaves grow older, however, they turn green. This silvery sheen, together with the fact that the tree (in contrast with another species, *Protea multibracteata,* the only other large tree-species in the Drakensberg) is a particularly graceful shape, is one of the best ways of identifying it. These leaves grow in clusters, or whorls, at the ends of twigs.

The flower-head is about 80 mm in diameter, the bracts as long as, or longer than, the perianth – another point of contrast with *Protea multibracteata,* as we shall see. They are red in colour, spoon-shaped, silky-hairy on the outside, and curve over backwards at the tip.

The flowerets are white below, but covered with pink to crimson hairs at the top. The style is white with a red stigma. Pollination is usually by birds, the Malachite Sunbird being a particularly happy and joyous visitor to this lovely flower. Another frequent visitor is a small, metallic-blue beetle, *Melyris natalensis,* and you will often find the flower-heads infested with this little chap.

The tree flowers from mid-summer to early winter. It grows on the slopes of the Little Berg, and is particularly abundant in the vicinity of cave sandstone cliffs. Its common name is the Transvaal Silverleaf.

The Manyika of Southern Rhodesia use *Protea roupelliae* with other plants to facilitate conception in women, and sometimes as a good-luck charm.

It was named after Mrs Arabella Roupell, a lady of the Cape, who published a beautifully illustrated book on South African wild flowers early in the 19th century.

Protea multibracteata is the only other large tree Protea in the Drakensberg, but it is very different from *Protea roupelliae.* Here the leaves are dark green and smooth, the bark black and crusted, and the tree itself is somewhat untidy, lacking the graceful shape of *Protea roupelliae.* The bracts are oblong (as against the spoon-shaped bracts of *P. roupelliae),* and glabrous, curving inwards this time to enfold the flowerets. The most striking feature of the 'flower', however, differentiating it clearly from *P. roupelliae,* is that the stamens are now much longer than the surrounding bracts, giving the 'flowers' a very typical spiky appearance. The flower-heads are about the same size and shape as *P. roupelliae,* rounded and saucer-like, and there are two colour varieties, a bright crimson and a cream.

Protea multibracteata is very similar to a Transvaal

Protea roupelliae

species, *Protea caffra.* In fact at one time our Drakensberg species was thought to be *P. caffra.* It was only fairly recently discovered that what was thought to be *Protea caffra* was actually *Protea multibracteata. Protea caffra* does not occur in the Drakensberg. (It was *Protea caffra* that appeared on the reverse sides of our old tickey and six-penny coins).

Protea subvestita is much smaller than either *P. roupelliae* or *Protea multibracteata,* and is more of a bush than a tree. It grows to a height of from 1,75 to 2,5 metres. It is also found at higher altitudes than the other two, about 2 130 metres, and in contrast to *P. roupelliae* and *P. multibracteata,* it lacks their spreading shape, being more columnar in form. The bushes also grow more closely together than in the case of the other two.

As in *P. multibracteata,* the bracts are oblong, but in contrast to this species they are hairy (pubescent) and curve in a pronounced way outwards, as the two pictures on page 103 show, away from the flowers, making the flower-heads bell-shaped rather than rounded. There are two colour varieties, a rose-pink and a cream.

The leaves and stems are slightly woolly, and the leaves distinctly veined. The woolly appearance gives the plant its specific name, for *subvestita* means partly-clothed. Living at high altitudes, in a rigorous climate, this Protea has donned woollen clothing for protection.

The first two Proteas we dealt with, particularly *P. multibracteata,* have thick, and in the latter case, gnarled bark, and this acts as an insulating feature and as a protection against fire. *Protea subvestita* lacks this thick bark, with the result that it is more liable to damage by fire, and is thus rarer than the other two. It is also found only in sheltered positions.

These three species are large and tree-like, if not actually trees. The other two species are quite different – dwarf-like and only a few centimetres in height.

Protea dracomontana is only about 250 mm high. The flower-heads, a rich salmon-pink in colour, have rounded bracts, and the leaves are broader than in any of the species already dealt with. They are bluish-green in colour, and are stiff and fleshy. A number of stems grow up from a single root-stock.

Like *Protea roupelliae* and *P. multibracteata,* this

Protea multibracteata

species is also highly adapted to withstand fires, but in this case the plant does it by developing a particularly strong root system, which sends up weak shoots readily renewed from below after fires. Like *Protea subvestita, Protea dracomontana* grows at a somewhat higher altitude to the other two, usually above the cave sandstone, from 2 000 to 2 300 metres. It flowers during February and March.

Protea simplex is a low shrub, similar in shape and size to *Protea dracomontana,* but instead of the rather bushy character of *Protea dracomontana* this species throws up only single, unbranched stems *(simplex* means unbranched), each stem crowned by a single flower-head. The colour is generally a pale cream, but there is a shell-pink variety. In shape, colour and general appearance the actual flower-heads are very similar to *Protea multibracteata.* Its range and habitat are also identical, except that it does have a tendency to grow at slightly higher altitudes to *Protea multibracteata.* There is actually a very close relationship between these two species. Like *P. dracomontana* it has a thick, underground rootstock. Buds start developing in December, and flowers mature in February.

One unique feature of *Protea simplex* is that it is almost entirely deciduous. All last season's shoots dry off and die, during the cold, dry winter, until spring comes and new stems emerge from the root-stock. The plant seems to benefit from veld fires.

Strangely enough, *Protea simplex* does not occur in the Cathedral Peak area, though it does occur elsewhere in the Drakensberg, both north and south of the Cathedral area.

It is sometimes known as the Sugar Bush, or Suikerbos, because of its sweet juice. When concentrated this is used as a useful cough mixture. (Actually the name Sugar Bush applies more correctly to *Protea repens,* not found in the Drakensberg). The dried shoots and leaves are used by the Xhosas to make snuff for the treatment of headaches and fever.

Protea simplex, in the Eastern Cape, is now an endangered species, owing to the encroachment of housing developments on its natural habitat, but in the Drakensberg it is fairly common. I know of one magnificent colony just below Blind Man's Corner, in the Cathkin area, only a few kilometres from my own home.

The Zulu name for all Proteas would appear to be isi-Qalaba.

Note: Just before going to press (August 1978) news was received of the discovery of a sixth Drakensberg Protea, *Protea nubigena,* growing on an almost inaccessible rock ledge on the Policeman's Helmet ridge, in the Royal Natal National Park. Nubigena, in Latin, means cloud-borne, or living in the clouds.

1

2

3

4

1. *Protea subvestita.*
2. Ditto: cream variety.
3. *Protea simplex.*
4. *Protea dracomontana.*
5. *Protea simplex,* rose-pink variety.

5

PHYTOLACCACEAE

This family is a small one, with only about 16 genera and about 100 species. The plants are natives of the Mediterranean region, the Canary Islands, central Asia, South America, Australia, parts of India, and to a limited extent South Africa, where we have only 7 genera and 31 species. Several species found in South Africa have been introduced from other countries.

PHYTOLACCA

Phytolaccaceae

There are three species of the genus *Phytolacca* in Natal, but as far as I know, only one of these, *Phytolacca octandra,* occurs in the Drakensberg, where it is common on disturbed land and on roadsides in the Little Berg. The plant is a shrub, rather straggly, about a metre high, usually found at about the 1 500 metre level. The inflorescence is a spike, about 50 mm long, bearing masses of closely compacted, very small, pale cream flowers. Each flower has a light green blotch at the base of the petals, there are five of these petals, and the flower has eight stamens. The plant flowers sparsely throughout the year, but autumn is the real flowering period. The pale cream flowers usually appear first in April. These soon turn to a spectacular mass of bright green berries, which in turn change to a glistening, brilliant blue-black colour. Finally, after the black berries are over, the calyx lobes turn to a sparkling crimson. All four stages are sometimes found at the same time on the same plant, making a striking show of colour. The stems are reddish, up to 0,5 m long, and branched. *Phytolacca octandra* is not indigenous to South Africa. It is an introduced species, coming originally from either Mexico or Japan.

The generic name comes from the Greek word *phuton,* meaning a plant, and Lacca, a place in Italy. *Octandra,* the specific name, comes from the Greek *okto* meaning eight, and *andron,* male, this referring to the eight stamens. The other two species in the genus that come from Natal are *P. heptandra* and *P. dodecandra.*

The Xhosas use an infusion from the root, which is large and resembles a sweet potato, to treat cattle suffering from lung sickness. The plant is also used by witchdoctors as an emetic. But some authorities claim that the plant is poisonous, and that it should certainly be regarded as poisonous to cattle. Watt and Breyer-Brandwyk report an African male as having died a few hours after taking a decoction of the plant as a medicine.

1

2

3

1. Persistent calyx of *Phytolacca octandra* after flowering.
2. Flowers of *Phytolacca octandra*.
3. Young plant of *Phytolacca octandra*.
4. *Phytolacca octandra,* showing the successive stages of green and black berries, and the persistent red calyx.

MESEMBRYANTHEMACEAE

The various genera in this huge family used to be included in an even larger family, that of the Aizoaceae, but a few years ago a large group of genera was taken out of Aizoaceae and given a family of their own, that of the *Mesembryanthemaceae,* the well-known 'Mesems'. It is extremely difficult to give a clear picture of this large family of succulents. Some years ago Herman Jacobson published a monumental work of three volumes on the Succulent Plants of the World, and the third volume, consisting of 525 pages and 463 figures, was devoted entirely to *Mesembryanthemum* and its allies. Even so, an enormous amount of work still has to be done on the various genera. They merge imperceptibly into each other, making identification extremely difficult.

The family is mainly South African and Mediterranean. The South African genera are found mostly in the arid regions of the south-west Cape, Namaqualand and South-West Africa. In Natal we have only four genera. All these are succulents, with fleshy leaves for the storage of water, and most have brilliantly coloured flowers. The strange mimic plants, *Lithops,* so beloved by collectors, are members of this family. The name of the family comes from two Greek words, *mesembria,* meaning midday (many of the flowers in this family only open in bright sunlight), and *anthos,* a flower.

DELOSPERMA

Mesembryanthemaceae

Delosperma obtusum is a low-growing succulent with brilliant scarlet flowers. It can usually be found growing amongst the sandstone rocks in the Little Berg, and it flowers from November to February. Altogether there are 25 species of *Delosperma* in Natal. *Obtusus* is a Latin word meaning blunted, dull or weak.

Professor W. P. U. Jackson, in his recent book "Wild Flowers of Table Mountain", tells of the fascinating way in which many of the plants in this genus ensure the dispersal of their seed. The seeds are contained in hard, dry capsules, but when these capsules become moist they split open, closing again when they dry out. When they split open the seeds are ejected, actually by the raindrops themselves, landing a foot or so away from the parent plant. This is done gradually: not all the seeds are ejected at the same time, some remaining in their capsules for a considerable period, even up to months and even years. "If we consider these facts", says Professor Jackson, "in relation to the natural arid habitat of most mesems we see how marvellously adapted they are. Thus their seeds are shed only in wet conditions and the spread in time, of both release and germination of seed, ensures that if the first rains are only slight and the early seedlings dry out, there is plenty of seed left for later. The proximity of the new seedlings to the parent plant renders it likely that conditions will be similar for all, and if suitable for the mother then they should be suitable for the offspring."

Delosperma obtusum

RANUNCULACEAE

The family of the Ranunculaceae is one of the most widespread, being found in most countries of the world. Altogether there are something like 48 genera and close on 2 000 species. In South Africa, however, its representation is small. In Natal we have only 5 genera and 9 species. Even so, the family contains some well-known plants – the Anemone, Clematis, and the Butter-cup being three of them. Four of the five Natal genera are found in the Drakensberg.

CLEMATIS

Ranunculaceae

One day in early October 1811, William Burchell, the famous South African traveller, camped beside a tributary of the Orange River, near Klaarwater. He came across a flower which gave a strange lift to his heart. It was one of the few flowers he had come across in his travels which reminded him, in a wild and savage country, of the flowers he had known and loved so well in England, and he made a special note of it. A month later, at Grootfontein, he found another specimen. This was *Clematis brachiata.* To-day we know that this attractive creeper is widespread throughout South Africa. Its common name is Traveller's Joy, and it was given this name because it is found in profusion all along the sides of the roads in the Eastern Cape. Other common names are Bridal Wreath and Klimop.

It is a woody climber with masses of creamy-yellow, sweet-scented flowers cascading from the bushes over which it climbs. The flowers have five pointed petals and a puff of yellow stamens in the centre. The small tapering leaflets are always arranged in threes. The flowers appear throughout the summer, and in the autumn the hairy, whitish seed-heads are equally prominent. In Natal it is common in the thorn country and along the margins of forests in the Drakensberg. *Clematis brachiata* is the only species of *Clematis* that we have in Natal. Elsewhere in South Africa there are only two species altogether.

The generic name *Clematis* comes from the Greek word *kleima,* meaning a vine. *Brachiata* means spreading out at right angles. The Zulu name is ihlozoleziduli. The Zulus pound the leaves and mix them with red earth, and this is smeared on children suffering from skin eruptions. The Xhosas have a tradition that a good show of flowers in early autumn heralds a good spring. They use the bruised stem as a snuff to clear the head.

The type specimen, *Clematis oweniae,* has an interesting history. For many years it was listed as occurring in Natal (both Medley Wood, 1907, and Bews, 1921, do this), and it was recorded by the famous botanist, W. H. Harvey, as having been collected at Port Natal. But Harvey and the other botanists were wrong. What had happened was this. It was named after Miss Owen, sister to the Rev. Francis Owen, missionary to the Royal Kraal of Dingaan in 1838, at the time of the massacre of Piet Retief. After the massacre Owen was driven out, and a year later he and his sister travelled through the Transvaal to Matabeleland, and it is thought that she collected the type specimen on this journey. But it was later despatched to England from Port Natal, and apparently Harvey must have assumed that it had been collected there, and it was thus recorded. *C. oweniae* does not occur in Natal.

Clematis brachiata

One of the most beautiful flowers in the Drakensberg in the early spring is *Anemone fanninii.* It is a white 'flower' with a yellow centre, usually found growing in moist depressions and gullies in the Little Berg. The 'petals' (strictly speaking they are not petals, but petaloid sepals) are creamy white with a silky texture on the outside, and the plant grows to a height of a metre or more. The leaves only grow to full height after the plant has flowered. They are rotund, and up to 200 mm in diameter, forming a compact growth. The 'flowers' are about 60 mm in diameter.

Over 150 known species of *Anemone* have been listed, but South Africa can lay claim to only three of them. In Natal we have two species, *Anemone fanninii,* which is common in the Drakensberg, and *Anemone caffra,* a Zululand species, but which I am told may occur in the Drakensberg. I have not seen it myself.

The word Anemone means wind-flower. It comes from the Greek word *anemos,* meaning wind, though just what the connection is, is not clear. (All I know is that for years I tried to get a decent picture of the flower, only to be beaten by the wind every time! The picture opposite was taken on a rare crystal-clear spring morning, after a night of rain, with the dew on the grass and not a breath of wind stirring, a morning which I shall long remember).

The specific name honours Mr George Fox Fannin (d.1865), a member of the gifted Natal family so well known in legal circles, and which has given such distinguished service to Natal for over a hundred years. George Fox and his wife were two of the earliest plant collectors in Natal. The Fannin name is also perpetuated in several other Natal flowers, notably *Disperis fanniniae,* so common in our Drakensberg forests, *Sisyanthus fanniniae, Fanninia caloglossa,* with its pale green, drooping flowers, and *Disa fanniniae.*

The Zulu name is uManzamnyama, which means black water. The Zulus make an infusion from the roots, which turns black. This is used as a purgative and also as a cure for gall-sickness.

Anemone fanninii

There has been considerable name-changing over the years in South Africa in the case of this genus, and not a little confusion in the matter of names as our knowledge of the genus has increased. To-day the position has been clarified, many of the older names have been dropped, and only three species are now recognised in South Africa. These three species are *Ranunculus baurii, R. multifidus* and *R. meyeri.* All three occur in the Drakensberg.

Ranunculus baurii (Ranunculus cooperi used to be included in this species, but the name *cooperi* has now fallen away) usually grows in the spray of waterfalls, from 1 800 to 2 500 metres. It is also found on stream banks in the Little Berg. The leaves are round (orbicular), and variable in size from 25 mm to 300 mm in diameter. In general it will be found that the higher the altitude the smaller the leaves become. On the summit the maximum diameter is 50 mm. These leaves have a finely crenate margin, and they are always thick and succulent. Flowers are yellow, and the plant flowers from December to March.

Ranunculus multifidus is an extremely variable species. It now includes the old *R. pubescens* and *R. pinnatus.* All forms of *Ranunculus* with dissected or deeply-lobed leaves are now classified under *Ranunculus multifidus.* It grows on stream banks and on the slopes and summits of the Little Berg. Flowers again are yellow, but much smaller than in *R. baurii.* The word *multifidus* comes from the Latin *-fidus,* meaning cleft. The common name is the Butter-cup Flower.

Ranunculus meyeri is a delicate plant with small, simple heart-shaped leaves, crenate on the margin, and with slender runners. It is often found growing in pools on the summit plateau. Like all other *Ranunculi,* the flowers are yellow.

Ranunculus multifidus is widely used as a medicine in South Africa. The Zulus powder the plant and use it as a cough remedy. The Xhosas drink a weak decoction for vomiting, diarrhoea and dyspepsia. The bruised leaf is used externally for cancer, scabies and mumps. The Zulus make a poultice from the leaves for the relief of colds and for venereal sores. Europeans in the Louis Trichardt district mix the ash of the plant with beef or mutton and apply this to the head of an infant suffering from fever. The Zulu name for the plant is u-Xhaphozi.

The name Ranunculus comes from the Latin word *ranum,* meaning a frog. The plants usually grow in damp places where frogs hop about.

1

2

1. *Ranunculus baurii.*
2. *Ranunculus multifidus.*

PAPAVERACEAE

Papaveraceae is the well-known Poppy family. It is a small family, confined mostly to the northern hemisphere, and flourishing in the temperate regions. There are only about 26 genera, with something like 200 species. Here in South Africa we have two genera, with only three species at most. All the plants in this family have a milky latex which possesses narcotic properties. The Corn Poppies of the English countryside and of "Flanders' fields", and our popular Iceland Poppies, are both members of this family.

PAPAVER

Papaveraceae

Clematis brachiata was not the only South African flower to awaken nostalgic memories in the mind of William Burchell (see page 108). On September 16, 1811, again while he was on the banks of the Orange River, we find him writing in the account of his Travels which he later published: "I found a Poppy four feet in height, with a showy bright red flower, like that of our common English Corn Poppy, an interesting and unexpected discovery in these southern latitudes of a genus so decidedly northern". He gave it the name *Papaver gariepinum* (in those days the Orange River was called the Gariep), but this was later changed to *Papaver aculeatum*. This is the only species of wild Poppy that we have in South Africa.

Actually Burchell appears to have been a little wrong about the colour. It is better described as a bright orange, sometimes a scarlet orange. The petals are crinkled, and there is a puff of yellow stamens at the centre. It usually grows to a height of about 0,75 m and leaves and stem are covered with yellow bristles. It flowers in midsummer. Although widespread throughout South Africa, it only appears sparsely, and in the Drakensberg only in the lower reaches.

The name Papaver is one that goes back something like 4 000 years. It is said to arise from the Celtic *papa*, meaning pap, a soft food for children in which seeds of *Papaver* had been boiled to make the infant sleep. *Aculeatum* comes from the Latin *aculeus*, a sting, thorn or prickle.

One peculiarity of the plant is that the buds hang downwards, not so much because of their weight but because of more rapid growth on one side of the stalk.

The plant used in the opium trade is, of course, a Poppy. This is *Papaver somniferum*. The narcotic properties in all the plants of the Papaveraceae family are particularly strong in this genus, and you will recall the "poppied sleep" of the poets.

Papaver aculeatum

CRASSULACEAE

Crassulaceae is a fairly wide-spread family, occurring in the temperate regions of Europe, Asia, South Africa and North America, but rare in South America and Australia. There are about 16 genera in the family, with over 400 species. The plants are almost always fleshy and succulent, and they are often found in dry, rocky places. The succulents of South-West Africa and the Namib desert are well known to horticulturists. The plants often make most useful rockery plants. They are able to survive in arid conditions because the plants store water in their thick, fleshy leaves. The name comes from the Latin word *crassus,* meaning thick.

In the Drakensberg we have three genera, *Cotyledon, Crassula* and *Kalanchoe.* All three are easy to differentiate by the amateur. In *Crassula* the stamens are the same in number as the sepals or petals (usually three to five). In *Cotyledon* the stamens are twice as many as the sepals or petals. These are always five in number. *Kalanchoe* is the same as for *Cotyledon,* except that the sepals or petals are four in number.

CRASSULA

Crassulaceae

Crassula is a large genus, no less than 55 species being listed as coming from Natal. At least 15 of these occur in the Drakensberg.

Crassula vaginata, illustrated on the opposite page, is a fairly common species in the Drakensberg, up to 300 mm tall. Each plant has a thick stalk with broad, fleshy leaves. The flowers are, individually, very small, but hundreds of these, massed together on a flattish circular head about 120 mm in diameter and usually a brilliant yellow in colour, make a striking flower-head. Sometimes paler yellow, almost cream, flowers are found, as may be seen in the pictures opposite. It grows on the lower slopes of the Little Berg, and flowers from January to March. The word vaginata comes from the Latin word *vagina,* a scabbard or sheath, and botanically refers to the sheath formed round the stem by the base of a leaf.

Crassula setulosa var. basuticus is a small red and white *Crassula,* only a few centimetres in height. The plants are characterised by rather coarse, hard hairs, and they form cushions on the rock faces from about 1 800 metres to the summit plateau. They flower in January and February. There is another variety, *var. curta,* which has pure white flowers. *Setulosa* comes from the Latin word *setosus,* meaning full of coarse hairs or bristles.

Crassula recurva is a rare *Crassula,* and I have an idea that it may occur only in the Drakensberg. I have seen it once, on the south-eastern slopes of Giant's Castle, at an altitude of about 2 200 metres. It usually grows on north-facing slopes, amongst the basalt boulders. The flowers are small, and like most Crassulas, closely-packed. Both red and white flowers may be seen growing on the same flower-head. The plant is about 250 mm high, the leaves acute, dark green, and streaked with purplish-brown patches.

The Zulu name for *Crassula vaginata* is u-Dumbukaye. The Zulus soak the leaves in hot water and use them to foment a bruise or sprained muscle. They also sometimes mix the ground-up root in their *amasi,* or sour milk.

1

2

3

4

1. & 2. *Crassula vaginata.* Note the two different colours.
3 & 4. *Crassula recurva.*

There are not many winter flowers in the Drakensberg, but in the dry and rocky gullies in the lower reaches of the Little Berg you will sometimes find an unusual flower called *Kalanchoe thyrsiflora*. It belongs to the genus *Kalanchoe,* a genus well adapted to dry conditions. As far as I know *K. thyrsiflora* is the only species of the genus that occurs in the Drakensberg.

The plant is usually found in rocky outcrops. It consists of a single, robust, fleshy stem which grows rapidly to a height of 0,6 m. The leaves, about 150 mm long and 70 mm wide at the base, are blue-grey in colour, and become progressively smaller as the plant gains height. The stem is four-angled, and both stem and leaves are covered with a silvery-grey powder, slightly sticky and resinous to the touch. Each stem ends with a head of small, closely compacted yellow flowers, which appear in June.

The plant is found throughout the Eastern Province, Natal, Zululand and up to the northern Transvaal.

It is supposed to be harmful to sheep. (Many of the plants of Crassulaceae are poisonous). The Sothos make a snuff out of the dried powdered roots, and this is used as a cure for colds. The Southern Sotho use a decoction of the plant as an enema against intestinal worms, but great care is needed, as it can be harmful. A decoction of the root is drunk by pregnant South Sotho women in cases of morning sickness, and it is used by some Africans as a charm to smooth away difficulties. The Southern Sotho name for the plant is seredile.

The common name of the plant is White Lady. *Kalanchoe* is the Chinese word for the plant, and means a four-petalled flower, while *thyrsiflora* comes from two Greek words, *thyrsus* meaning a stalk or stem, and *flora,* a flower.

Kalanchoe thyrsiflora

ROSACEAE

This is the Rose family, and one of the most valuable plant groups that we have, for it contains some of our loveliest cultivated flowers, and also most of our fruit trees. The apple, pear, plum, quince, peach, apricot, cherry and strawberry all belong to this family. Most of the plants are shrubs or trees. Distribution is worldwide, but mainly in the northern hemisphere. There are about 100 genera and 2 000 species.

LEUCOSIDEA *Rosaceae*

There is only one species in this genus, *Leucosidea sericea,* and it is endemic to South Africa. This is the well-known *umtshitshi,* a densely-leaved shrub or small tree, which is usually found along stream beds at altitudes between 1 250 and 2 200 metres. In the Drakensberg it is the only decent-sized tree above the 2 000 metre level, and is a good friend to the mountaineer, both for its welcome shade and for its use as firewood. It is found in the northern Transvaal, Natal, the eastern Orange Free State, Lesotho and in the eastern Cape from Cathcart to Queenstown. In the spring it comes out in a mass of small yellow flowers, arranged in a short terminal spike about 40 mm long.

Umtshitshi is the Zulu name for the tree. Its common name is Oudehout or Ouhout. The generic name comes from the Greek word *leukos,* meaning white and probably the Latin *sidus,* starry, and hence beautiful, bright or shining. *Sericea* comes from the Latin word *sericus,* silken, which in turn comes from the Latin name for China, *Seres,* noted for its silken fabrics.

RUBUS *Rosaceae*

The genus *Rubus* includes all the well-known blackberries, raspberries, brambles, loganberries and so on. The name Rubus is the old Latin name for a bramble. It is a huge genus – there are almost three thousand described species – and they occur in most parts of the world. Natal can claim only six of these, and as far as I know only three occur in the Drakensberg.

Rubus cuneifolius is the well-known Wild Blackberry. You will know all about it if you ever get tangled up in it in one of the valleys of the Little Berg! It has canes of up to two metres in length, equipped with vicious, curved prickles. It grows at about 1 750 metres, often at the base of sandstone cliffs. The fruit is green, turning to red and then purplish-black. It is edible, but sour. The flowers usually come out in November.

Rubus ludwigii is a small shrub, smaller than *R. cuneifolius,* with prickly stems. The leaves are green above and whitish below. It is often found growing under Protea trees, and flowers during November and December. The berries are crushed by the more literate Africans, water is added, and the resultant liquid, red in colour, is used as ink. The Zulu name for the plant is i-Tshalo.

Rubus pinnatus is found occasionally in *Podocarpus* forests. The fruit is bright orange in colour, and rather sweet to the taste. The Xhosa name is amakonuble, and hence comes the name of the well-known coast resort near East London, Gonubie, where this species is prolific.

1

2

3

4

1 & 2. *Leucosidea sericea.*
3. *Fruit of Rubus cuneifolius.*
4. *Rubus cuneifolius.*
5. *Rubus ludwigii.*

5

LEGUMINOSAE

The Leguminosae family is the third largest family of flowering plants in the world. It contains close on 600 genera and over 12 000 species, and is distributed throughout the world, though it is somewhat rare in New Zealand. In Natal we have 89 genera, with 420 species. The family includes such well-known plants and trees as the Acacia (this includes all our thorn-trees), the Mimosa, Bauhinia, Lupins, Sweet-peas, Cassia, Wisteria, Erythrina, and such useful plants as peas and beans, lucerne and clover. Many of our South African timber trees belong to this family. One distinguishing feature of all the plants in this family is that the fruits have bean and pea-like pods.

Apart from their many commercial uses, leguminous plants always improve the quality of the soil. This is because the roots have growths on them which contain a bacterium, Rhizobium, which converts nitrogen from the atmosphere into a nitrate which is needed for plant growth.

BUCHENROEDERA *Leguminosae*

Buchenroedera is mainly a South African genus, with about 20 species altogether, but there is one species that comes from tropical Africa. In South Africa the genus is confined to the eastern areas, from Natal to Uitenhage. We illustrate one species, *Buchenroedera lotononoides*.

This is a small bush, about 0,45 m high, with bluish-green leaves, usually found growing on the slopes of the Little Berg at about the 1 600 metre level. It has small blue flowers, which, like all the flowers in this family, are pea-shaped. It blooms throughout December and January.

There are two other species of *Buchenroedera* in the Drakensberg, *Buchenroedera amajubica,* with white or pale blue flowers, usually found growing at about the 1 850 metre level, and *B. tenuifolia var. pulchella,* which is found at a much higher altitude, amongst the basalt cliffs, round about the 2 500 metre level. It is often found in the upper reaches of the passes leading to the summit. *Buchenroedera amajubica* is somewhat rare.

The word lotononoides means resembling Lotononis, another plant which is a member of this family. Lotononis itself is a combination of the generic names of two plants, Lotus and Ononis.

Buchenroedera lotononoides

SUTHERLANDIA *Leguminosae*

Sutherlandia montana is easily recognised. It is a shrubby bush growing to a height of a metre and more, and bearing masses of brilliant scarlet pea-shaped flowers, which droop down in clusters. The flowers are about 30 mm long, and they are followed by large green pods. The plant is often found growing in the dried boulder beds of streams in the Little Berg. I once found a magnificent specimen in the bed of the Mnweni River, near its confluence with the Ifidi River, and the photograph opposite (bottom left) was taken in the bed of the Umlambonja River, in the Cathedral area. It likes a dry, sandy soil, and is common in the Karroo. The stems are stiff and reedy, and the plant flowers from early spring to May.

Common names are Gansies Keur, Balloon-pea and Kankerbos. It was once thought by the early Dutch settlers to be a cure for cancer, hence the last name. The Zulu name is u(lu)-Nwele. It is used by the Zulus as a blood purifier and as a tonic, and also for chest colds.

The genus was named in honour of James Sutherland (c.1650-1719), one of the earliest Superintendents of the Edinburgh Botanic Gardens. (He must not be confused with Dr P. C. Sutherland, who gave his name to *Greyia sutherlandii,* the Natal Bottle-brush. See page 134).

Altogether there are six species to the genus, which is endemic to South Africa. Natal can claim only two of these, *Sutherlandia montana* and *S. frutescens*. The latter, as far as I know, does not occur in the Drakensberg, though *S. montana* is often confused with it. The two plants are very similar.

ERIOSEMA *Leguminosae*

Eriosema kraussianum is a small yellow flower, about 150 mm high, growing on the grassy slopes of the Little Berg at about the 1 400 metre level, and flowering in early spring. It grows in small clumps. Altogether there are 11 species of *Eriosema* in Natal, four of which are found in the Drakensberg. The genus is widespread, being found in tropical Africa, America, Australia and southern Asia. In South Africa it extends from South-West Africa, the Transvaal, Natal, Lesotho and along the eastern coast regions to Humansdorp and Knysna.

In the Drakensberg *Eriosema* plants appear in early spring, in grassland which has been burnt. They often flower and fruit even before the first spring rains, while the grasses, which cannot commence their spring flush before the rains, are still dormant or short. In comparison with their size they possess particularly strong and deep-penetrating root systems, which put them into contact with soil water too deep-seated for the grasses.

The only medicinal use for *Eriosema* that I know of in South Africa is the use of *Eriosema salignum* by the Sotho as a stimulant for bulls, but elsewhere in Africa, particularly in West Africa, Senegal, Zaire and northern Nigeria species of *Eriosema* are used extensively as medicine. In Zambia *Eriosema psoralioides* is used in the treatment of syphilis.

---

1 & 2. *Sutherlandia montana.*
3. *Eriosema kraussianum.*

1

2

3

GERANIACEAE

The Geraniaceae family is another that is widespread throughout the world, especially in temperate and sub-tropical regions. There are five genera and over 800 species. Our Geraniums and Pelargoniums are members of this family, many of the world's finest hybrids having come originally from endemic South African plants. Many species are aromatic, and some are still used medicinally, both overseas and by African witch-doctors. The family name comes from the Latin word *geranos,* a crane. This is because the fruit resembles a crane's bill or beak. We have four genera in South Africa, of which three, *Geranium, Pelargonium* and *Monsonia* are represented in the Drakensberg.

MONSONIA

Geraniaceae

On a day early in 1774 three people, a lady and two gentlemen, might have been seen making their way across the Cape Flats and heading for the Darling district, north-west of Cape Town. The two men were well-known Cape botanists, Carl Peter Thunberg and Francis Masson, while the lady was Lady Anne Monson, great-granddaughter of Charles II. Ten years earlier the elder Linnaeus had written her a long and rather passionate 'love' letter, claiming to have been infatuated with her for a long time, although he had never met her. He asked her whether he could name a newly-discovered Cape flower after her, "so that your name may be commemorated in the Kingdom of Flora". She, of course, agreed. In 1774 her husband, Col. George Monson, had been appointed a member of the Indian Council, and she was on her way out to join him in India. Darling was the home of the flower named after her, where to this day it makes glorious splashes of colour to delight the heart of the tourist, and Lady Anne was on her way to make the acquaintance of 'her' flower. She stayed for some little time at the Cape, and the three became firm friends. Before she finally left she gave her name to another South African flower, the pure white Erica, *Erica monsoniana,* growing in the Franschoek area.

Monsonia flowers all have five 'petals', and the genus is easily recognised if one examines the stamens. There are 15 of these altogether, and they are arranged in five groups of three. The actual species, however (there are about 40 altogether), are difficult to recognise, as they merge imperceptibly into each other.

As far as I know, only one species occurs in the Drakensberg, and this is *Monsonia attenuata.* This is a small white flower, about 30 mm in diameter, with wavy-edged petals, white in colour and with a yellow centre and long narrow leaves. The backs of the 'petals' are distinctly veined, and the plant is only about 100 mm in height. It flowers during January and February, and grows at about the 1 500 metre level.

Two other species of this genus *M. biflora* (renamed recently as *Monsonia angustifolia)* and *M. ovata* are well-known as excellent and proved remedies for dysentery and diarrhoea, and also for insomnia. Round about 1890, in response to a request from England, Dr J. Medley Wood, the well-known Natal botanist, and Curator of the Durban Botanical Gardens, collected a sackful of the latter species at Charlestown, and sent them over to India for use in the military hospitals there.

Monsonia attenuata

OXALIDACEAE

The plants in this family used to be classified under *Geraniaceae,* but many years ago they were taken out and placed in a family of their own. The family is widespread, occurring in Africa, Madagascar, North and South America, Asia and the Malay Archipelago. There are seven genera and something like 900 species. In South Africa it occurs mainly in the Cape. In Natal we have only one genus with 5 species. All the South African species in this family have small nutlets, or bulbs, a very uncommon feature amongst dicotyledons. All the flowers have five slightly overlapping petals, with a characteristic twist.

OXALIS

Oxalidaceae

The genus *Oxalis,* the Wild Sorrel, is well-known, especially to gardeners struggling to eradicate the clover-leafed weed that is such a pest in Natal gardens, *Oxalis corniculata.* In most species the plants are small, about 150 mm high. The leaves are like three-leaved clovers, and the plants have small, deep-seated, nut-like bulbs which, if disturbed, multiply rapidly and are difficult to eradicate. Some species have a special device for scattering the seed over a wide area. Petals are five and rounded, overlapping, and usually pink to red in colour, but in the Cape, where they flourish, colours range from white to pink, yellow, orange and mauve. In dull weather, even if it is hot, the petals close up tightly, a device, apparently, to keep the pollen dry.

The plants contain Oxalic acid, which is a poison. (The name Oxalic comes from two Greek words, *oxys,* acid and *als,* salt). The leaves are ideal for the cleaning of brass or copper, and for removing ink stains from clothing.

There are about 500 species of *Oxalis,* mostly confined to South Africa, but they are also found in South America and in Mexico. In South Africa they are mainly Cape plants, flowering in the winter and dying down in summer. In Natal, however, where we have five species, they flower in midsummer.

The common Drakensberg species, with its bright pink flowers, is *Oxalis obliquifolia.* It is found almost anywhere in the Drakensberg, from 1 200 to 2 500 metres, except on the summit of the Main Berg.

Another Drakensberg species is *Oxalis smithiana.* It grows at a height of about 2 000 metres, and flowers in October and November. Unlike *Oxalis obliquifolia,* it does not grow singly, but in small groups. The leaves, too are more slender.

The Xhosa name is inkolwane. The Xhosas dry and powder the nutlets as a remedy for tapeworm.

1

2

1. *Oxalis obliquifolia.*
2. *Oxalis smithiana.*

POLYGALACEAE

The Polygalaceae family has 12 genera and over 800 species, distributed world-wide in all warm and temperate countries except New Zealand. In Natal we have only two genera and 35 species. Both genera occur in the Drakensberg. We illustrate here one of these, *Polygala*. Superficially the flowers resemble those of the pea family, Leguminosae, but are quite different. There are three petals in Polygalaceae, more or less attached to the stamens, the front petal keel-shaped. All three are two-lipped. In addition, the fruit is not a pod, as in the pea family, but usually a two-celled capsule. The European name for the family is Milkwort. This is because some of the species were supposed to increase the milk yield in cows. No reports of increased milk production, however, have been received from South Africa. The name Polygala comes from the Latin *poly* meaning much or many, and *gala*, milk.

POLYGALA

Polygalaceae

The genus is a large one with about 600 species found throughout the world. Nine species occur in the Drakensberg, but here we illustrate only one, *Polygala virgata*.

This is a twiggy shrub, about 1,5 metres tall, usually found growing on the edges of bush in the Little Berg and alongside stream banks. The branches bear drooping sprays of reddish-purple flowers. Each flower is enclosed by two large purple bonnet-like bracts, which are streaked with darker veins. These open to reveal the flower which has a purple tuft of tiny hairs at the top of the lower keeled petal. The two outer petals enclose the lowest petal like a bonnet. The purple tuft of hairs is typical of all Polygalas, and is the salient feature in identifying the various species. The plant flowers in mid-summer.

The word virgata comes from a Latin word meaning twiggy. Common names are Purple Broom and Bloukappies, while the Zulu name is u-Sihlonishwayo. Many of the plants in this family are used medicinally by the Zulus, particularly as remedies for tuberculosis and dropsy. An infusion is made from the roots of *Polygala virgata* and used to *phalaza* with to purify the blood.

Polygala virgata occurs both in Tropical and East Africa, southwards through the Transvaal and Natal, and into the Cape as far as George. In the Drakensberg it is found between 1 250 and 1 800 metres.

Polygala virgata

MELIANTHACEAE

Melianthaceae is a small family, consisting of only two genera and 15 species. These genera used to be included in the Sapindaceae family, but in 1850 Engler established the separate family of the Melianthaceae. Up to quite recently it included the genus *Greyia* (see next page), but this genus is now in a separate family of its own. The Melianthaceae family is found in most tropical countries and in South Africa.

MELIANTHUS

Melianthaceae

Melianthus villosus, the Honey Flower, is a small bush from 1,5 to 2,0 m tall, with deeply indented and wavy leaves. These leaves, when crushed, are somewhat unpleasantly scented, and this explains the other common name of the flower, Kruidjie-roer-my-nie (Gertie, touch-me-not). The genus has six species in South Africa, and is widespread in both dry and humid areas.

The word Melianthus comes from two Greek words, *meli* meaning honey, and *anthos,* a flower. The flower secretes masses of sweet nectar, hence its common name. *Villosus,* in Latin, means hairy, and a glance at the first picture opposite will show the aptness of this specific name.

The flowers of many species of *Melianthus,* such as *Melianthus insignis,* are a bright red. Our Drakensberg species, *M. villosus,* is rather drab in appearance. It usually grows below the sandstone cliffs at an altitude of about 1 800 m, and flowers in midsummer up to about the end of February. The flowers, about 30 mm long, are green in colour and heavily streaked with brown. They hang downwards on the stem in a closely-compacted pendulous raceme, which may be up to 300 mm long.

The structure of the individual flowers is most intricate and very interesting. There are five sepals of varying widths and shapes. Hidden by them, but sometimes partly showing, are four small, tongue-like petals. Two of these are a bright red in colour, not showing in the third picture opposite, but clearly seen in the first picture. These two red petals and two of the sepals, long, narrow and pointed, no doubt act as guides to the pollinating insects or night-flying moths which visit the flower to get at the nectar.

One of the sepals at the base forms a deep, cup-like receptacle in which the nectar is held. This is partly enclosed by the other sepals which form a hood over it to protect the honey from dew or raindrops. Hidden also in the sepals is the small, rounded ovary and style, surrounded by four stamens in two pairs. During the development of the stamens and anthers the style and stigma remain hidden among the anther filaments, an adaptation, no doubt, to guard against the chances of self-pollination and increase the chances of cross-pollination. They emerge only after the stamens have withered.

One of the most interesting things about the flower is its development. At first the nectar-holding receptacle is uppermost, but at an early stage in its development the pedicel, or stem, twists through 180 degrees, turning the flower virtually upside-down and bringing the receptacle into the right position for holding the honey.

The whole structure of the flower and its development is a marvellous piece of work, and beautifully adapted for pollinating, either by insects or, I suspect, by night-flying moths "that drink them under the moon".

Melianthus villosus. The top picture, greatly enlarged, shows a single flower before it makes its 180° *turn*.

GREYIACEAE

Greyiaceae, a small family of only one genus and three species, used to be classified under Melianthaceae, but it has recently been given a family of its own. The smallness of the family is made up for by the spectacular beauty of the flowers. They are a brilliant, gay scarlet, borne on small woody trees or shrubs. The one genus, *Greyia,* is essentially a mountain one, and is indigenous to South Africa. It is found nowhere else.

GREYIA

Greyiaceae

The scarlet flame of *Greyia sutherlandii,* with the blackish-grey bark of its trunk and its bright green leaves shining in the sun against the blue skies of early spring, will remain long in your memory if you visit the Drakensberg in August or September. *Greyia sutherlandii* is a small tree, about two or three metres high, and it grows just above and below the rock bands of the cave sandstone. The spectacular flowers, a brilliant red, form a cylindrical mass of small, closely-packed, cup-shaped flowers with protruding red stamens. Each cluster of flowers is about 70 mm in diameter, and the flowers themselves are bi-sexual. Its common names are Mountain Bottlebrush and Baakhout, and the Zulu name is i-Ndalu. The Zulus make an infusion from the roots, and this is supposed to be a cure for biliousness.

There are two other species in the genus, *Greyia flanaganii* and *Greyia radlkoferi.* All three species are found only in the eastern parts of South Africa, and nowhere else in the world. The three species are fairly easily differentiated.

The leaves of *G. sutherlandii* are devoid of hairs (glabrous) and the petals are oblong. In *G. radlkoferi* the leaves are densely covered with short, soft hairs (tomentose) giving them a soft, velvety feel. This easily distinguishes this species from the other two. Also, in *G. radlkoferi* the petals narrow slightly at the base. *G. radlkoferi* is found only in the Transvaal and in northern Natal, while *G. sutherlandii* grows mostly in the Drakensberg, Swaziland and the Orange Free State, though it does occur occasionally in the Transvaal. In *G. flanaganii* the flower spikes are shorter and not so freely produced as in the other two. The leaves also are a lighter green. This species comes from the Queenstown area.

The name *Greyia sutherlandii* commemorates two well-known historical characters, Sir George Grey, Governor of the Cape from 1854 to 1861, and Dr P. C. Sutherland, Surveyor-General of Natal from 1856 to 1887, one of Natal's most colourful characters and a noted plant collector. (For further details on this remarkable man, see page 142).

The plant was first collected by Dr P. C. Sutherland in 1859, but it had already attracted the attention of Captain Allen Gardiner 23 years earlier. Writing in his "Travels in the Zooloo Country" in 1836, Gardiner said: "While toiling up a very steep ascent this afternoon" (in the Underberg district) "I observed a remarkable tree, about the size of a large apple tree, growing near the bank. From the circumstances of its bearing a reddish blossom, and being totally divested of leaves, I first mistook it for the Umseensi" ('um-Sinsi' is the Zulu name for the Kaffirboom, *Erythrina caffra* or *E. lysistemon)* "but on nearer inspection I found that the blossom was a deep cherry or blood-red colour, and am inclined to believe that it is a species little known."

Greyia sutherlandii

TILIACEAE

This is the Lime, Linden or Jute family. It consists mainly of trees and shrubs, and altogether there are something like 50 genera and 450 species, fairly widespread throughout the world, especially in the tropics. Only four genera occur in South Africa, and as far as I know, only two of these, *Grewia* and *Sparmannia,* are found in the Drakensberg, each with one species. Several species of plants in this family are used for their fibre especially the Jute plant, *Corchorus olitorius* (found in Tongaland) and *C. capsularis.*

SPARMANNIA *Tiliaceae*

Sparmannia ricinocarpa is a rather lovely white flower, growing on a spreading shrub often over a metre tall. It is usually found near water, often in the shade of stream-bank scrub or in moist depressions. The flower has eight petals and a mass of striking yellow stamens. The flowers are about 20 mm in diameter.

Sparmannia is an African genus of seven species, with a few in Madagascar, but only two are found in South Africa, and only one of these, *Sparmannia ricinocarpa,* occurs in the Drakensberg. The other species, *S. africana,* is a shrub growing up to seven metres in height, with flowers 40 mm across. This species is found only in the Cape, from Knysna to Port Elizabeth.

The genus was named after Andrew Sparrman, the famous Swedish botanist and traveller who spent from 1772 to 1776 (except for an 18 month voyage with Captain Cook) botanizing in the Cape. The specific name, *ricinocarpa,* comes from the Latin word *ricinium,* meaning a mantle with a hood or veil attached, and another Latin word, *carpus,* a seed.

MALVACEAE

The Malvaceae family is found throughout the world in tropical and temperate regions. Altogether there are about 75 genera and close on 1 000 species. It is the Hibiscus and Hollyhock family, and the cotton plant, *Gossypium herbaceum,* also belongs to it. In Natal we have 13 genera, three of which are found in the Drakensberg. The flowers are frequently very showy, all with five petals. They are easily recognised, for the filaments of the stamens are united in a tube surrounding the pistil, the style protruding through the top of the tube. The word Malvaceae comes from the Greek word *malva,* and refers to the salve-making properties of many of the plants in this family.

Sparmannia ricinocarpa

The various hybrids of *Hibiscus* make popular cultivated flowers, especially the brilliant red species so well-known on the Natal coast, *Hibiscus rosa-sinensis.* In fact, the number of hybrids would fill a good-sized book. In their wild state there are something like 300 species, but only about 50 of these are found in South Africa.

The genus is more of a coast and tropical one than a mountain one, but at least two species have been identified in the Drakensberg, *Hibiscus trionum* and *H. aethiopicus.*

Hibiscus trionum, illustrated opposite, is a small, straggly shrub, found growing in grasslands in the valleys of the Little Berg. It has five delicate pale yellow petals, with deep purplish-maroon centres. The stems are reddish-brown and the dark green leaves slender, lobed and hairy. The filaments of the stamens are united into the usual characteristic column-like tube, with the style protruding through the top of the tube. The anthers are a bright yellow. The plant flowers in December and January. Common names are the Bladder Hibiscus (referring to the shape of the fruit) and Black-eyed Susan. It is widespread throughout South Africa, occurring in all four Provinces, and often flowering along roadsides. The Xhosas use a decoction made from the roots of this plant as a remedy for red worms.

Hibiscus aethiopicus is somewhat similar to *H. trionum,* but is easily distinguished from the latter. The flowers lack the dark centres of *H. trionum,* and the anthers are brown in colour.

The word Hibiscus comes from the Greek name for the Marshmallow. In the early 1960's efforts were made in the sub-tropical parts of the Transvaal to establish a fibre industry from the plants of *Hibiscus cannabinus,* but little came of this.

The Sotho name for *Hibiscus trionum* is solwane. The Xhosas use the plant as a dressing for septic wounds.

Hibiscus trionum

CLUSIACEAE

This is an alternative name for the Guttiferae family. It is not a particularly large family, there being only 49 genera and 1 200 species. Only two of these 49 genera appear in South Africa, *Hypericum* and *Garcinia*, from which comes our gamboge. Only *Hypericum* appears in the Drakensberg, *Garcinia* being a coast genus. The family is fairly widespread, in all warm and temperate countries.

HYPERICUM

Clusiaceae

The *Hypericum* genus with 400 species is easily recognised. All the flowers are yellow, and each flower has a tuft of long, spiky-looking stamens at the centre, which are unmistakable. It is mainly a northern-hemisphere genus. The well-known St. John's Wort is one of the species. There are only seven species in South Africa, of which four occur in Natal and three in the Drakensberg.

Hypericum aethiopicum subsp. sonderi is easily recognised as a species from the black or reddish-brown glandular dots on the leaves and flowers, those on the margins of the petals being particularly noticeable. The flowers are the usual bright yellow, with the typical tuft of central stamens. The plant is small and bushy, about 150 mm high, and it flowers from November to January. It is fairly common, and is also found in the Orange Free State, the Transvaal, Moçambique and Rhodesia. The plants are poisonous, but the Zulus use a decoction made from the roots as an enema to ease backache and pains in the loin. They call it the uSukumbili. *H. aethiopicum* is used by Southern Sotho women for administration to young girls at the time of puberty.

The derivation of the name Hypericum is uncertain. It could be related to a Mediterranean plant. Alternatively, Hyperion, in Latin, is one of the names for the sun, and it is possible the name derives from this word, for the brilliant yellow flowers are certainly sun-like.

BEGONIACEAE

This is the Begonia family, with five genera and about 900 species. It is found In America, Asia and Africa. South Africa has only one genus, *Begonia,* with 8 species in Natal at the moment, though the genus is at present undergoing revision, and this number may be reduced. Only one of these, *Begonia sutherlandii,* occurs in the Drakensberg, as far as I know. Most Begonias are monoecious, the male flowers having four petals and numerous stamens, and the female five petals usually, and three styles. The leaves are often highly decorative, being strongly veined and of different colours.

Hypericum aethiopicum subsp. sonderi

Most Begonias come from moist, tropical countries, particularly South America. As we have seen, there is only one genus in South Africa, *Begonia,* with 8 species from Natal. Only one of these occurs in the Drakensberg, *Begonia sutherlandii.*

This is one of the most attractive of our South African Begonias. It is widespread in the Drakensberg, and grows in damp, cool places in our *Podocarpus* (yellowwood) forests, usually on the south faces of moist rocks, (often in company with the purple *Streptocarpus gardenii),* and on the sides of streams. The flowers are monoecious, and the usual pattern of Begoniaceae flowers prevails, male flowers with four perianth segments, and female with five. The leaves are light green and taper to a point, and they have serrated edges and red veins. The stems also are red.

Begonia sutherlandii was named after Dr P. C. Sutherland, one of the outstanding characters in early Natal history. He was born in Scotland in 1822, and early in his life developed an insatiable curiosity for all things scientific. As a young man he qualified as a doctor, graduating in medicine at the Royal College of Surgeons in Edinburgh. He then visited West Africa, travelled to Greenland, and, still only 23 years old, took part in an expedition to try and find Sir John Franklin, who was missing in the Arctic. After writing a two-volume account of this expedition, and having an island in the Arctic named after him, he decided to emigrate to Natal, and was one of the first medical men to come out to this country. Thereafter he threw himself heart and soul into the affairs of the infant colony — Surveyor-General, noted plant-collector, Chairman of the Harbour Board, official Colonial Geologist (he discovered the Dwyka Series near Durban), founder of the Pietermaritzburg Botanical Society, and moving spirit in the youthful Natal Society, still flourishing to this day, and still running the well-known Natal Society Library. He was a member of the Legislative Council for many years, and his travels on horseback took him into all parts of Natal, giving him a wide knowledge of its geological and botanical features. As we have already seen, the well-known Natal Bottle-brush, *Greyia sutherlandii,* was also named after him.

Begonia is named in honour of Michael Begon, a Frenchman (1638-1710) one time Governor of French Canada and an enthusiastic plant-collector.

The Zulus, who call the plant u-Qhamamaveni, use it to find out whether a kraal is threatened in any way by an enemy. The whole plant is first of all ground to a powder, which is then sprinkled on half a calabash, and the grinding stone is then placed on top of it. This process is repeated until there are three such stones on the calabash. If it withstands the weight, no enemy threatens, but if the calabash topples over, disaster will befall the kraal!

1

2

3

1. *Begonia sutherlandii* growing in the shade of the deep forests of the Little Berg.
2. *Begonia sutherlandii* showing female flowers.
3. Male flower of *Begonia sutherlandii*.

THYMELAEACEAE

Thymelaeaceae is the Daphne family. World-wide, there are about 40 genera in this family, with roughly 400 species, and they occur in Africa, the Mediterranean region, Australia, and a few in Asia and North and South America. South Africa has only 13 genera, 7 of which occur in the Drakensberg. The plants are mainly trees and shrubs, though a few herbs appear. The flowers of many species grow in clusters. Many of the larger species have a tough, fibrous bark, used by the Hottentots and early settlers as rope. Those well-known flowers of our gardens, Veronica, Hebe and Daphne are all members of this family.

LASIOSIPHON

Thymelaeaceae

This genus and the next, *Gnidia,* are very closely allied, and the differences between the two genera are slight and often ill-defined. Recently there have been several transpositions between the two genera. At present, 12 species of *Lasiosiphon* are recognised from Natal, and four of these occur in the Drakensberg.

Lasiosiphon polyanthus is a shrubby plant, up to a metre in height, with masses of brilliant yellow flower-heads. It grows at about the 1 750 metre level, and flowers in summer. It is said to be poisonous to stock. The Zulu name is i-Nhlanhla elimpofu. An infusion is made from the roots and used by young men when out courting. The leaves are also used medicinally. They are crushed and rubbed on the face as a cure for pimples. Dr Andrew Smith, (1797-1872), the great South African naturalist and traveller, was of the opinion that the roots of many of the Lasiosiphons could usefully be used to make lotions for the relief of sore throats. The word Lasiosiphon comes from two Greek words, *lasiois,* meaning shaggy (both leaves and stems of these plants are often clothed with shaggy hairs) and *siphon,* a tube.

GNIDIA

Thymelaeaceae

Gnidia capitata is one of the plants that has recently been transferred from *Lasiosiphon* to *Gnidia.* Its old name was *Lasiosiphon capitatus.* It will be found growing in open grasslands in the lower reaches of the Little Berg, during December and January. It is a slender plant, about 300 mm high, with a rounded head of small yellow flowers. (The word capitata comes from the Latin *capitatus,* meaning a head). Each flower has a slender calyx tube and five sepals. The petals are very small. It is reputed to be highly poisonous, but the Zulus, who call it Umfukuzane, use it as a cure for snake-bite and as a remedy for sore throats. Its common names are Kerrieblom and Gifbossie.

The origin of the name Gnidia is uncertain. It is, possibly, a Greek word for Daphne, or Laurel, or it could be connected with the Greek town of Knossos in Crete.

---

1. *Gnidia capitata.*
2 & 3. *Lasiosiphon polyanthus.*
4. *Gnidia capitata.*

1

2

3

4

Dais cotinifolia is the only species in this genus in South Africa. There is one other species, indigenous to Madagascar. *D. cotinifolia* is a small tree which bears heads of small, pinkish-mauve flowers throughout December and January. Its common name is the Pompon Tree, and the Zulu name is i-Ntozwane emnyama, which means black fibre plant. There are a number of fibre plants whose bark is used for tying bundles of wood or thatch. *Dais cotinifolia* is recognised as supplying the best and strongest fibre of them all. This has given rise to another of its common names, Kannabas. The bark tends to turn black when it dries.

The tree usually grows to a height of about 4 metres. At first its shape is tall and narrow, but it soon begins to round out with age. The bark is grey, and the leaves, light green at first, soon turn to a rich glossy dark colour. Veins and midribs are a pale lemon-yellow. The flower-heads measure about 40 mm across, and consist of a number of tiny flowers clustered together. Each flower has a long, thin curved hairy tube which opens into five lobes. The name Dais comes from the fact that each flower is placed on a little platform or 'dais', made by the bracts. These remain on the tree long after the flowers have fallen.

It is interesting to note that this tree, though not indigenous to the Cape, was known to botanists soon after 1760. In 1776 it was introduced to both the Continent and England. Linnaeus was the first to describe it in Europe, in 1764, from plants brought over from South Africa by sailors. Its tidy shape and glossy leaves, quite apart from its attractive flowers, make it a welcome tree in many parts of the world. In South Africa it is found in its wild state along the eastern seaboard, from East London to Natal, the eastern Transvaal, and up to the Soutpansberg.

Dais cotinifolia

ERICACEAE

This is the well-known Heath family, with a world-wide distribution, especially in the temperate zones. In the tropics the plants are usually found growing at the higher altitudes. Altogether there are something like 50 genera and more than 1 350 species. There are 8 genera in South Africa, with only two in Natal, *Erica* and *Philippia*. The plants in this family are mainly trees or shrubs, many of them with the characteristic small, narrow, needle-like leaves, although in the northern hemisphere plants such as the well-known Azaleas, Rhododendrons and Bilberries belong to the family.

ERICA

Ericaceae

Erica is essentially a mountain genus, and is well represented in the Drakensberg, where there are at least 20 different species. Ericas are common in Europe, all the way from Norway to the Mediterranean and from the Azores to Turkey, including, of course, Scotland, where the mountain heather has inspired poets from the earliest times. (Actually, the term 'heather' is best reserved for the British *Calluna vulgaris* rather than *Erica.)* But in all this vast area there are only 14 species. A few species occur in the mountains of North Africa and East Africa, but it is when one reaches the Cape that the full glory of this lovely genus is encountered. In South Africa as a whole there are 612 species, and no less than 95 per cent of these occur in the mountains of the south-western and southern Cape. Nowhere else in the world, possibly, is there such a concentration of species of a single genus. Another remarkable fact is that many of the Cape species have a relatively small range, some as little as the size of a rugby field, and occur nowhere else. One exception to this is *Erica cerinthoides,* which is found from the Cape right up into the northern Transvaal and on into Rhodesia.

All Ericas are woody plants, but they range in size from tiny plants only a few centimetres high to fairly large trees like *Erica caffra,* with a trunk 0,3 m in diameter, and *Erica arborea,* a North African species which grows to a height of 6 metres. So large are the roots of this Erica that briar pipes are made from them, the very name 'briar' coming from the French word for Erica, bruyere.

Ericas are easily recognised from their leaves. These are short and needle-like, and usually arranged in little tufts, clustered very densely. The flowers are small, sometimes only the size of a pin-head, but again, they are closely massed, and often make a cloud of glorious colour. The individual flowers are shaped like little bells, long curved tubes, or puffed tubes.

The name Erica comes from the Greek, *ereike,* a word used by Pliny and Theophrastus to describe the genus. This in turn comes from the Greek word *ericein,* meaning to break or pound. This refers either to the supposed property of some species to break gall-stones, or because the stems are brittle and break easily.

Probably the best known of all the species of *Erica,* although it is not particularly common in the Drakensberg, is *Erica cerinthoides.* This is partly because of its striking colour, a brilliant, sparkling red, but also because of its wide distribution and the fact that it flowers throughout the year. Common names are Rooihartjie and the Red Hairy Erica. The flowers have puffed scarlet tubes, covered with downy red hairs, and the tubes are slightly constricted at the mouth. It is named after the genus *Cerinthe,* (the honey-wort), the suffix -oides meaning similar to, or like. It is common throughout South Africa, and on up into Rhodesia. It was introduced into cultivation in England by Francis Masson in 1774. The prevailing colour is red, but there is a white variety that comes from Swaziland. In the Drakensberg there is a somewhat rare variety where the colour is a pale golden-salmon. I have only once come across this variety, on a summer's day many years ago, in the Ndedema Valley. Fortunately I had my camera with me, and the result may be seen on the opposite page.

Erica algida is a small Erica, 150 to 250 mm high, with a brownish-pink flower, the flowering period being midsummer. It is usually found on the summit of the Little Berg, up to the 2 500 metre level.

Another small Erica, only about 200 mm high, is *Erica alopecurus.* It is usually found growing in damp places near small mountain streams, and flowers from March to May. The flowers are pink in colour and are crowded together into a cylindrical spike, making this species easily recognisable. It is this feature which

Erica cerinthoides, the lower picture showing the rare pale golden-salmon variety.

gives the species its name, for *alopecurus* means, in Greek, a fox's tail.

Erica caffrorum is a white Erica, growing to a height of about 0,65 m. It grows in the boulder beds of streams, and flowers in September and October. Occasionally a pink variety is found.

Erica drakensbergensis is a tall, bushy plant, from 0,3 to 1,2 m tall, with small, white campanulate flowers and dark brown anthers. It flowers from January to July, and is fairly widespread, being found in the northern part of the eastern Cape, Natal, Lesotho, and up into the Soutpansberg in the northern Transvaal. With its pure white flowers and bright green leaves, it makes a most pleasing sight in the many rock gullies one finds at the summit of the Little Berg.

Erica drakensbergensis and *E. caffrorum,* both white, are very similar to each other, and are often confused. In *E. drakensbergensis* the tube is more rounded, with the throat constricted and slightly toothed. The ovary and the anthers are brown, and, especially the anthers, giving a marked brown 'centre' to the flower. In *E. caffrorum* the tube is more elongated and the brown ovary and anthers not so prominent. The flowering times also are different, *E. drakensbergensis* flowering in March and April, while *E. caffrorum* flowers in September and October.

On the summit of the Little Berg I recently came across an Erica that was new to me. It had reddish-brown flowers, and was growing right on the edge of the plateau, on a south-facing slope, at an altitude of about 2 000 metres. The time was late March. It was identified as possibly a new species, but also possibly a low-altitude form of *Erica dominans.*

Erica maesta is a browny-red Erica, about 0,8 m high. It flowers in September, and may be found on the summit of the Little Berg at 2 000 metres, though it sometimes grows at a lower altitude.

One of the commoner Drakensberg Ericas is *Erica oatesii.* It flowers from March to August, and has pink, bell-shaped flowers, somewhat larger than many of the Drakensberg Ericas, while the anthers have distinct curved spurs. It grows to a height of about 400 mm, like so many of our Drakensberg Ericas on the summit of the Little Berg. It is widely distributed, appearing in the Transvaal, Swaziland, the Orange Free State and Lesotho, in addition to Natal and the Drakensberg. It

Erica Cerinthoides

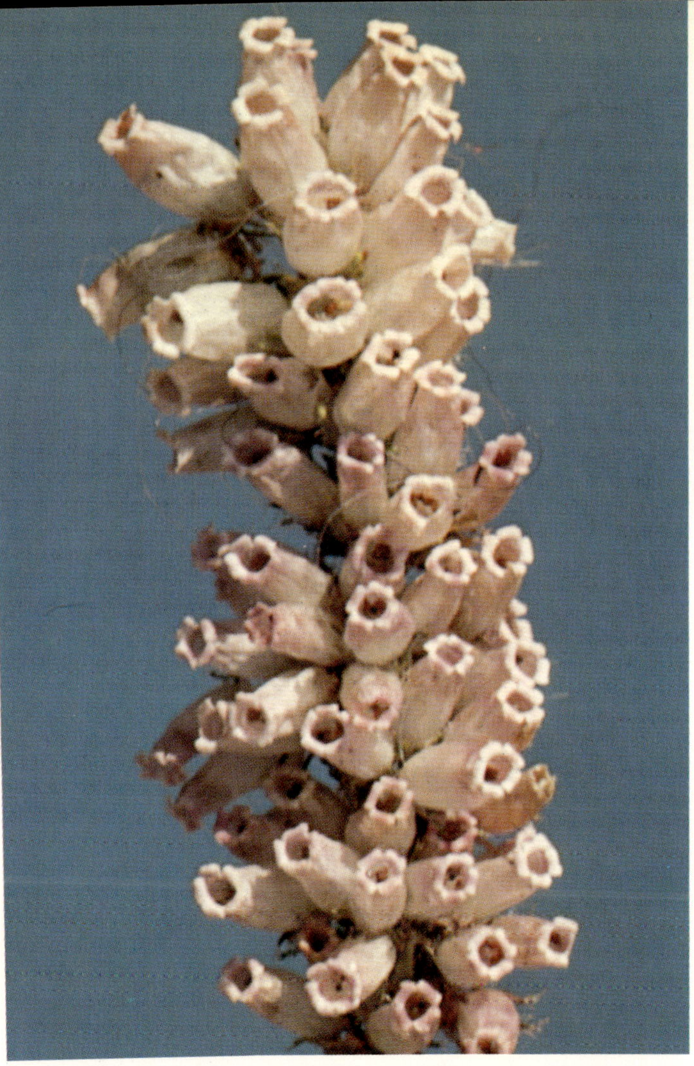
1

3

4

2

1. *Erica alopecurus.*
2. *Erica oatesii.*
3. Low-altitude form of *Erica dominans,* or possible new species.
4. Close-up of latter.
5. *Erica drakensbergensis.*

was named after Frank Oates, the English traveller and naturalist, who collected the species in 1873.

Erica schlechteri is one of the larger Ericas in the Drakensberg, growing to a height of nearly a metre. Its flowers are pink, and they appear from August to November. It is usually found in the higher altitudes, on rocky, basalt outcrops.

Erica thodei was named in honour of Hans Justus Thode, (1859-1932), who spent a good deal of time in the 1890's collecting plants "in the majestic range of the Kathlamba (sic) or Drakensberg" to quote the man himself, "which has irresistibly attracted and again and again induced me to climb its lofty heights . . . the goal of my long-cherished dreams." *Erica thodei* is a small plant, only about 300 mm high, found on the summit plateau at 3 100 metres, where it forms mats against the rocks. The flowers vary in colour from white to pink, and the flowering period is from January to March. Distribution is from Barkly East to the Natal Drakensberg. Thode's name is commemorated in no less than thirteen South African plants, ten of which occur in the Drakensberg. These are *Alepidea thodei, Athanasia thodei, Disa thodei, Erica thodei, Holothrix thodei, Lessertia thodei, Manulea thodeana, Osteospermum thodei, Romulea thodei* and *Sebaea thodeana*.

Erica straussiana (it used to be called *E. westii)* is usually found growing on rocky stream banks and along the margins of forests in the Little Berg. It is a dwarf shrub, about 0,4 m high, and flowers in September. The flowers are pale pink to white, and the tubular petals are sharply pointed.

Erica woodii has small, red urn-shaped flowers, and the flowering period is March. It is about 300 mm high. It is a wide-spread species, occurring all the way from the eastern Cape, through the Transkei to Natal, Lesotho, Swaziland and the Transvaal. It was named in honour of J. Medley Wood, Curator of the Durban Botanic Gardens in the early years of this country, and one of Natal's best known plant collectors.

And finally, *Erica wyliei*. This is a creamy-white Erica, about 300 mm high, which usually grows in south-facing passes leading to the summit.

Ericas are particularly susceptible to fires, but this has already been dealt with in the Introduction.

1

2

3

4

1. *Erica woodii.*
3. *Erica caffrorum* var. *caffrorum.*

2. *Erica algida.*
4. *Erica straussiana.*

LOGANIACEAE

The Loganiaceae family, the Strychnine and Wild Sage family, is a small one, with about 30 genera and 400 species distributed world-wide, mainly in the tropical regions. Natal has four genera, and two of these occur in the Drakensberg, *Buddleia* and *Gomphostigma*. Strychnine, the virulent poison, is obtained from an Indian species of this family.

BUDDLEIA

Loganiaceae

In early spring the soft foam of the Buddleias is a common sight in the lower reaches of the Little Berg, while their fragrance is all-pervasive in the green valleys.

Buddleias are fairly large shrubs sometimes almost tree-like, and they are found growing wild in all four Provinces of the Republic. Six species are found in Natal, and at least three of these occur in the Drakensberg, *Buddleia auriculata var. euryfolia, B. corrugata* and *B. salviifolia.*

Perhaps the commonest of these is *Buddleia salviifolia.* Normally a fairly large shrub, it can grow into quite a fair-sized tree, as one of the pictures opposite shows. It flowers in late winter and early spring. The flowers are small and tubular, each tiny flower being about 5 mm in length, with a yellow-orange throat. Individually the flowers are unattractive, but they are borne in such close masses that they make a beautiful sight. There are two colour varieties, a creamy white and a pale lilac. They have a strong, sweet scent, and the leaves are long and tapering, and thickly felted on the undersides.

This genus is one of the first to establish itself on the edges of mountain forests after a fire, and is a precursor to the larger forest trees such as the yellow-wood, *Podocarpus latifolius,* and the white stinkwood, *Celtis africana.* In the early days the Zulus used the wood as shafts for their assegais. The Xhosa name is cwanci.

The genus was named in honour of Adam Buddle, an English botanist. Common names are Sage Wood, or Saliehout, the names probably derived from the fact that the leaves are rather salvia-like.

Buddleia auriculata is somewhat smaller than *B. salviifolia.* It has orange or salmon flowers, tipped with yellowish cream. The leaves are a little shorter than *B. salviifolia.*

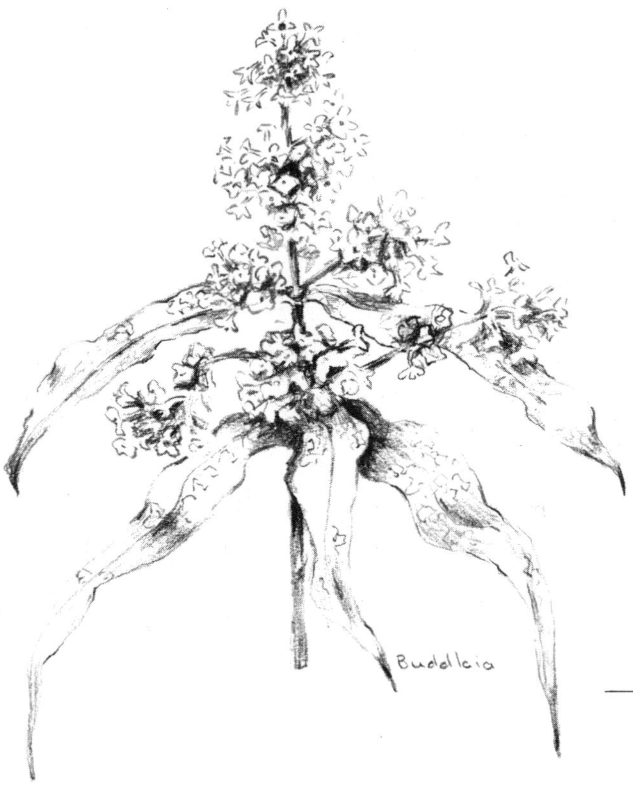

Buddleia salviifolia, showing both the lilac and the cream varieties.

GENTIANACEAE

Gentianaceae, the Gentian Family, is distributed worldwide in the temperate regions, and also on tropical mountains. Altogether there are about 64 genera in this family, with about 600 species. Natal can lay claim to six of these genera. Some of the plants are aquatic and semi-aquatic, and all have blue or yellow flowers. The Alps are well-known for their beautiful blue Gentians. Here we illustrate two of the best known genera in this family, *Sebaea* and *Chironia*.

SEBAEA

Gentianaceae

The genus is well represented in the Drakensberg, there being at least 11 different species in these mountain regions. The flowers are usually yellow in colour, but are occasionally white. They grow on the higher levels, on the summit of the Little Berg, and in a few cases right up to the summit plateau at 3 300 metres. The genus is widespread throughout South Africa, except in the central areas. It is also found sparingly in Madagascar, India, Australia and New Zealand.

The genus was named after Albert Seba (1665-1736), a Dutch naturalist and apothecary, who founded a famous museum known as "The Wonder of Europe".

Sebaea natalensis (up to recently it was called *Sebaea rudolfii),* is a small yellow flower, growing in clusters at the tips of short stems, its usual habitat being the grasslands of the Little Berg at about the 1 500 metre level, often near vleis. It is about 150 mm high, and the flower clusters are from 50 to 100 mm across.

Its Zulu name is isi-Vumelane. A warm-water infusion is made from the pounded roots and this is used by young men to *phalaza* with when courting.

In midsummer the bright yellow flowers of *Sebaea thodeana* will be found on the wet sides of waterfalls in the Little Berg, and along stream banks. It is a creeping perennial, and it forms dense mats of thickly-clustered flowers wherever it grows. It is found at all levels from the summit plateau down to 2 100 metres. It flowers from September to January. The flowers are clustered closely together, and are about 15 mm in diameter.

And finally we have *Sebaea sedoides var. sedoides.* Here the flowers are a butter-yellow, but sometimes white varieties are found. It is rather more robust and branched than the other species we have been considering. It grows on the slopes of the Little Berg, at about the 1 750 metre level, and flowers in early May.

1

2

3

4

5

1. *Sebaea natalensis.*
2. *Sebaea thodeana.*
3. *Sebaea sedoides var. sedoides.*
4. *Sebaea thodeana.*
5. *Sebaea sedoides var. sedoides.*

Climbing the slopes of the Little Berg, one day in midsummer, I came across a pink flower that was new to me. It was identified as *Chironia peglerae,* and a few weeks later I came across another species at 'The Lakes', in the Giant's Castle Game Reserve.

There are 25 species of the genus *Chironia* in South Africa. A few are found in Tropical Africa, and also in Madagascar. Otherwise it occurs nowhere else in the world. In South Africa it is mainly a coast flower, but two species, I discovered, are found in the Drakensberg. The flowers in this genus are almost always pink, rarely purple, and another characteristic of the genus is the tendency of the anthers to twist. The inflorescence is always a cyme. The two Drakensberg species are *Chironia peglerae* and *C. krebsii.*

Chironia peglerae stands about 0,2 m high. Like most of the Chironias the petals are pink, and in this species, pointed. The flower has five stamens, which are yellow and shorter than the petals. The flowers are about 20 mm in diameter. It is not a particularly common flower in the Drakensberg. It has been identified in the Ndedema Gorge and on the slopes of the Little Berg in the Cathedral area. The specimen illustrated opposite, which I photographed, was found on the slopes of the Little Berg in the Cathkin area, at an altitude of about 1 800 metres. The plant flowers in January and February.

Chironia krebsii, the other Drakensberg species, was named after George L. E. Krebs (1792-1844), a German naturalist from Hanover. The plant is usually found in marshy ground at high altitudes, where mists are frequent. It is common in the coast regions, from the Cape to Natal, and also in the Transvaal, Orange Free State and Swaziland. It flowers in December and January. Leaves are strap-shaped, and, like *C. peglerae,* the plant has a single stem. Flowers are pink, but occasionally a white variety is found. In appearance it is very similar to *Chironia peglerae.* This species is used by Southern Sotho women as a remedy for colic and diarrhoea in their children, and also by pregnant women.

The generic name, given by Linnaeus, comes from Chiron, one of the centaurs. He taught mankind the use of plants and medicinal herbs. He was also an astronomer. When he died there was no room for him in the northern skies, so Zeus put him in the southern skies, and his name was given to the two pointers to the Southern Cross, Alpha and Beta Centauri. (For many years Alpha Centauri held the distinction of being the nearest known star to our solar system, but recently a faint star, not very far from Alpha, called Proxima Centauri, has been discovered, and this was found to be a little nearer.)

1

2

1. *Chironia krebsii.*
2. *Chironia peglerae.*

ASCLEPIADACEAE

There are about 220 genera in this family, with over 1 800 species. The plants are usually found in tropical and sub-tropical regions, especially in Africa. Only a few genera are found in temperate regions. Natal has 31 genera, 12 of which occur in the Drakensberg. Many of the plants in this family are succulents, and the majority have milky or watery juice, which gives rise to the common name Milkweeds. Some are climbers, and the well-known Stapelias belong to this family. The flower structure is complicated and highly specialised to assist in insect pollination. The family name was given by Robert Brown (1810) after the Greek God of Medicine, Asklepios.

PACHYCARPUS *Asclepiadaceae*

Pachycarpus is an African genus of about 30 species, 26 of which are found in South Africa. It occurs in the eastern areas, from Swellendam northwards along the coast, and includes the eastern Orange Free State. It is rare in the south-western and central areas. 14 species occur in Natal.

Pachycarpus campanulatus is a weak-stemmed plant, about 0,5 m in height, found in the grasslands of the Little Berg, with greeny-yellow, bell-shaped flowers, hanging down in a cluster. Because of this habit I have always found it difficult to spot the flowers, as the bells are the same colour as the grass, and I have only rarely seen the flower. Each flower is about 20-40 mm in diameter, and they appear during January.

There is also a fairly common variety of *P. campanulatus* that occurs in the Drakensberg, *P. campanulatus var. sutherlandii.*

Here the flowers are smaller than those of *P. campanulatus*. We might also note a rare species, *P. plicatus,* recorded so far only from the Royal Natal National Park at Mont-aux-Sources.

The milky sap of *P. campanulatus* is sprinkled by the Zulus on eggs to prevent dogs from eating them. If a dog eats one of these eggs he is supposed to foam at the mouth and retch. Africans also rub the milky sap into the mouths of dogs to make them savage. The Zulu name for the plant is i(li)-Shongwe.

The name Pachycarpus comes from two Greek words, *pachus,* meaning thick, and *carpus,* a fruit. *Campanulatus* means bell-shaped. The various species of *Pachycarpus* are very similar to each other, and it is difficult to distinguish between them.

SCHIZOGLOSSUM *Asclepiadaceae*

Schizoglossum decipiens is a small flower, found growing on the lower slopes of the Little Berg in November and December. The flowers are greenish-yellow, with the perianth segments striped with brown lines. As they develop, however, the flowers turn a purplish-brown. Leaves are long and narrow, slightly hairy, and in pairs of whorls. The whole plant stands about 150 mm high, and the leaves are about 50 mm long. The stems have a milky juice.

As a genus, *Schizoglossum* is found mostly in the eastern portions of South Africa, in Natal, the Transvaal, Lesotho and the eastern Free State. There are about 90 species in South Africa, with 48 coming from Natal.

1

2

3

4

1. *Pachycarpus campanulatus.*
2. *Schizoglossum decipiens.*
3. *Pachycarpus campanulatus.*
4. *Schizoglossum decipiens.*

BORAGINACEAE

Boraginaceae is a family found in most parts of the northern hemisphere, where there are about 95 genera and 1 600 species. In South Africa we have only 16 genera. The flowers are borne mainly in cymes, and several of our garden flowers, such as Heliotrope, Delphinium, Anchusa and the Forget-me-nots belong to this family. Four genera occur in the Drakensberg. We illustrate two of them, *Tysonia* and *Myosotis*.

MYOSOTIS

Boraginaceae

No account of the wild flowers of the Drakensberg would be complete without a reference to this dainty flower, the wild Forget-me-not, *Myosotis sylvatica*. It is indeed a welcome sight to the tired climber as he breasts the last rise to the summit and knows that his climb is over. Blue is a somewhat rare colour amongst Drakensberg wild flowers. Perhaps nature thought that with the blue skies of these mountain regions, and the royal blue of the distant peaks, she should go sparingly in the use of this lovely colour.

Myosotis sylvatica is a slender, somewhat trailing plant, found growing in damp places and in the sheltered rock gullies on the summit plateau at 3 000 metres. It does sometimes occur, however, in the lower reaches. It flowers throughout the summer, and the flowers are small, only about 5 mm in diameter, a bright sky-blue in colour with yellow centres.

Myosotis is a small genus – there are only three Natal species, with a total of 50 species, fairly widespread throughout the world. The name comes from the Latin *mus,* a mouse, and *ous,* an ear. The Southern Sotho name for it is sethutu sa mathuela.

TYSONIA

Boraginaceae

There is only one species of *Tysonia, Tysonia africana,* and it is found only in Natal and East Griqualand. It is a smallish, straggly bush, with stiff stems, a metre and a half in height, with the flowers grouped in a large, lax terminal panicle of cymes. Leaves are long and lanceolate in shape, and larger at the base of the stem. Flowers are very small, and white in colour. The effect is that of a white, diaphanous cloud against the landscape.

The genus was named after Mr W. Tyson, one of Natal's early plant collectors, who collected from 1881 to 1890.

VERBENACEAE

This, the Verbena family, is a large one, found mainly in the tropical regions of both hemispheres. There are something like 73 genera and over 2 000 species. Only a few genera are found in temperate regions – Natal has 13. The different genera and species vary considerably in the appearance of the flowers. Well-known plants that are included in this family are our garden Verbenas, the Petrea with its shining blue cloud of flowers in the spring, the well-known Lantana, now declared a noxious weed, and Clerodendron. Teak comes from a tree which is a member of this family, *Tectona grandis*. All the plants in this family have tough, hard stems. The word verbena comes from the Latin word, *verber,* meaning a stick.

1

2

3

4

5

1. *Tysonia africana.*
2-4. *Myosotis sylvatica.*
5. *Tysonia africana,* greatly enlarged.

Verbena bonariensis is not a particularly showy flower, but it is widespread in the Drakensberg and deserves a place in this book. It is a stiff-stemmed plant, 1,25 m high, and the tips of the stems branch out into a series of smaller stems, each bearing a closely-compacted head of very small, purple flowers. The stems were often used by the Bushmen for arrow-shafts, and the flowers appear from December to March.

Verbena venosa is a much more showy flower. It is often found growing alongside the roads in the Drakensberg and on the lower slopes of the Little Berg, making a flat carpet of brilliant purple, about 100 mm high. There is also a white variety, not as common as the purple. Like so many other species in this genus it is an introduced species, coming probably from South America. The word verbena, as we have already seen, comes from the Latin word *verber,* meaning a stick, while *venosa,* in Latin, means full of veins.

Although widespread throughout the world, only six introduced species of *Verbena* occur in South Africa. The Drakensberg has three of these, the two we have already mentioned, and *Verbena brasiliensis.* The Sotho use a decoction of the root of *Verbena venosa* for heart-burn and colic. Their name for the plant is morod.

LABIATAE

This family, also sometimes called Lamiaceae, is another family that is found in warm and temperate regions of the world, especially in the Mediterranean countries. It is only rarely found in Arctic and Alpine regions. It is the Salvia, or Sage family. The stems are usually quadrangular (i.e. square in cross-section) and practically all species have salvia-like flowers, the five sepals joined to form a bell-shaped calyx, and the five petals also joined, with the corolla two-lipped. (The word labia means, in Latin, a lip). Many of the species are strongly aromatic, and a number of our garden herbs, such as lavender, rosemary, mint, sage and thyme belong to this family. Salvia, Leonotis and the well-loved Coleus also belong to it.

HEMIZYGIA *Labiatae*

Hemizygia elliottii is a small white flower, with the usual salvia-like characteristics, often found growing along stream banks and on the lower slopes of the Little Berg. It usually grows to a height of about 0,2 m, but in favourable conditions can sometimes reach 0,6 m. In this species the long yellow-tipped anthers against the translucent white of the petals are particularly prominent.

Altogether there are about 30 species of *Hemizygia,* of which 21 occur in South Africa. It is essentially a Transvaal genus, where one species, *Hemizygia obermeyerae,* is a shrub from two to four metres high. The word Hemizygia means capable of being divided into two sets of halves, in one plane only, and refers to the filaments. It flowers in midsummer. Four other species occur in the Drakensberg, *H. stenophylla, H. bolusii, H. pretoriae* and *H. teucriifolia.*

---

1 & 2. *Verbena bonariensis.*
3 & 4. *Verbena venosa.*
5. *Hemizygia elliottii.*
6. *Verbena venosa.*
7. *Hemizygia elliottii.*

1

2

3

4

5

6

7

PLECTRANTHUS

Labiatae

Note: *Plectranthus calycinus* has recently been taken out of *Plectranthus* and re-named *Rabdosia calycina.* As it is the only South African species in *Rabdosia,* it has been thought better in this instance to include it in its old genus, *Plectranthus.*

The Plectranthus genus consists of soft-wooded, evergreen shrubs with numerous spikes of blue, mauve, pink or yellow flowers, which are typically salvia-like. They are tubular and two-lipped, and arranged in a long, tapering spike.

Rabdosia calycina (the old *Plectranthus calycinus:* see note above) is a pale yellow flower, the specimen illustrated opposite having been photographed in the Setene Valley. Its common name is the Spur Flower, and it flowers from January to March.

The word Plectranthus comes from the Greek word *anthos,* a flower, and a Latin word *plectrum.* This was a small stick with which the player struck the chords of a stringed instrument, and refers, no doubt, to the shape of the plant. An alternative, and possibly more correct, derivation is from the Greek *plectron,* a spur. Some species have narrow tubular flowers like the spur on a bird's leg.

Three other species of *Plectranthus* which occur in the Drakensberg are *P. dolichopodus,* which grows in the forests of the Little Berg, *P. grallatus,* found in moist gullies and along stream banks at about the 1 500 metre level, and *P. fruticosus,* which grows on the slopes of the Little Berg. One other species which may occur in the Drakensberg is *P. nummularius,* popularly known as the Gossip Plectranthus, because of the rapidity with which it spreads!

An infusion is made by Africans from the roots of many species of this genus and used as a cure for colds and as a mouth-wash.

STACHYS

Labiatae

Stachys kuntzei is a small, white salvia-like flower which may be found on the slopes of the Little Berg, at about 1 750 metres, in the months of October and November. It is about 200 mm high, and the leaves, fairly small (30 mm long) are light green and heart-shaped. The flowers are also rather small and insignificant.

Stachys is a Greek word meaning a spike, or an ear of corn, this referring to the shape of the inflorescence.

1

2

3

1 & 2. *Rabdosia calycina (Plectranthus calycinus)*
3. *Stachys kuntzei.*

When the tranquil days of autumn come to the valleys of the Drakensberg, you will be charmed with the rich colours of the wild Leonotis. These lovely flowers, with their tall minarets of bright orange, make an unforgettable picture against the blue of the distant peaks as the cold days of winter approach.

Seven species of *Leonotis* have been recorded from Natal. Best known of all, and commonest in the Drakensberg, is *Leonotis leonurus*. It grows in dense clumps, with stiff, erect stalks which reach a height of over two metres. The leaves are narrow, with serrated margins, slightly hairy, soft and tapering, usually about 80 to 100 mm long. They grow thickly at the base and thin out at the top. The flowers, tubular in form, are a rich, bright orange, velvety in texture, and are borne in successive whorls at intervals along the tall, elegant stem. They have a strong, aromatic scent, and they make a brave show of colour in the valleys and along the slopes of the Little Berg.

It is a popular flower, and has been given many common names. Minaret Flower (very descriptive), Lion's Ear, and Lion's Tail are three of them. It has also been called Wild Dagga, because the leaves are dried and smoked by the Zulus, and strangely enough they are rather similar to the true (and notorious) Dagga plant, *Cannabis sativa*. The two plants, however, are quite unrelated, and *Leonotis leonurus* has no narcotic properties at all.

It is, however, used in a large variety of medicinal ways. It appears to possess purgative properties, and has been used in England in the composition of a patent medicine. The whole plant, except the root, is boiled down and used as a tonic for calves. The reputed dose, which is said to be very efficacious, is half a litre twice or three times a week. Both leaves and roots can be pounded together and immersed in the drinking water of sick fowls.

There are two Zulu names for the plant, u-Tshwala-bezinyoni omncane and i-Munyamunyane. The generic name comes from two Greek words, *leon,* a lion, and *ous,* an ear, referring to the shape of the flower. *Leonurus,* in Greek, means lion's tail. If the lion is the king of beasts, then indeed is this a kingly flower!

The various species of *Leonotis* are so closely allied that it is extremely difficult to tell the difference between them. Sometimes the difference depends on nothing more than the thickness of the flower clusters and the distance between the whorls.

Dr Andrew Smith (1797-1872), noted naturalist and explorer, who was an acute observer of the ways of nature, made a special note of this flower when he visited Natal in 1832. After remarking that the sugar-birds were particularly partial to the sweet nectar at the bottom of the long trumpet-shaped corollas, he says in his Journal: "Before the mouth of the corolla opens, which it does when the stamens are mature, the nectar is intensely bitter, but at the moment of opening the sweetness is developed. This means that nature does not wish insect marauders who cannot carry the pollen where it is required, to come and rob the nectary."

Leonotus leonurus was one of the first South African plants to be cultivated overseas. Jacob Breynius was growing them in Europe in 1663.

Leonotis leonurus, the top picture against the majestic backdrop of Cathkin Peak.

BECIUM *Labiatae*

On your climbs up the Little Berg you will sometimes come across a very attractive, dainty flower with salvia-like petals, white in colour. This could be *Becium obovatum*. The plant is about 180 mm high, and it flowers throughout early spring. Its common names are Cat's Whiskers and I Spy.

The inflorescence is a series of opposite cymes with dainty white frilly-edged flowers, sometimes tinged a faint pink. Style and stamens are very prominent, the filaments being long and twisted. The whole inflorescence has a whiskery-like appearance, which no doubt has given rise to one of its common names. The leaves are small, oval, and have a delicate perfume.

It is widely distributed, from the Eastern Cape, through the Transkei to Natal, the Free State, Eastern Transvaal, Swaziland and Rhodesia. The genus, actually, is a small one, with only about 10 species altogether, two in India and the rest in Africa. In South Africa there are only four species, and *B. obovatum* is the only one that occurs in Natal.

The Zulu name is i(li)-Dada. The Zulus make an infusion from the pounded roots, and this, when warmed, is used as an enema for children suffering from stomach troubles. It is also reputed to be used as a hair-restorer.

AJUGA *Labiatae*

Ajuga ophrydis is a small shrub, about 0,25 m high, bearing purplish-white flowers with the characteristic two-lipped corolla, on a slender spike. The tube of each flower is somewhat enlarged at the throat, and is slightly hairy, with four stamens. The leaves are short and toothed. It is often found growing on stream banks and in moist areas. It flowers from November to January, and usually grows at about the 1 500 metre level.

There are about 100 species of *Ajuga*. This is the only species to be found in South Africa, where it occurs in the eastern regions, from the Transvaal and eastern Orange Free State to Natal, and along the coast to Uitenhage and the Somerset East districts.

Amongst the southern Basotho the plant is burnt, the ash carefully collected and mixed with fat, and the mixture placed on pegs which are then driven in round a kraal. This is believed to protect the kraal from the entry of small-pox. The Basotho name for the plant is senyarela.

1

2

3

1. *Becium obovatum*.
2. *Ajuga ophrydis*, a close-up of the flower.
3. *Ajuga ophrydis*.

SOLANACEAE

The Solanum family is important commercially, for some of our best known vegetables belong to it. The potato is one, and also the tomato, *Lycopersicum*. The Egg-plant, the Cape Gooseberry and all our Red Peppers also belong to it. The Tobacco Plant, *Nicotiana*, belongs to this family, and from *Atropa* we obtain the alkaloid Atropine, used so extensively in medicine. Some of our finest garden flowers are also included in this family, such as our Petunias and Nierembergia. It contains something like 72 genera, and between 2 000 and 3 000 species, found in all tropical and temperate countries, especially in South America.

All species of Solanaceae contain the poisonous alkaloid solanine, but in some the amount is so small as to be harmless. A drop of juice from *Solanum sodomaeum* will turn milk sour. Solanum is the Latin name for the deadly Nightshade, also a member of this family.

SOLANUM

Solanaceae

Solanum aculeatissimum is well-known for its bright yellow rounded berries, 20 to 25 mm in diameter, growing on spreading bushes about 0,5 m high. Common names are Bitterappel, Devil's Apple and Apple of Sodom. (The Apple of Sodom was a mythical fruit, very beautiful externally, but when grasped, turning to smoke and ashes). The flowers are white with a central tuft of yellow stamens, very like the flowers of the potato plant. The root is sometimes used as a remedy for toothache, and an ointment made from the burnt plant is supposed to be helpful in cases of rheumatism, but the berries themselves are highly poisonous.

Solanum nigrum is the well-known um-Sobo, the black berries of which make such delicious pies and jam. The plant is much smaller than *S. aculeatissimum*, only about 0,3 m tall, and the flowers, which are white, are also small, about 5 mm across. The strange thing about this plant is that the berries, when eaten raw or half ripe, are highly toxic, but are harmless otherwise. Common names are Black Nightshade and Galbessie. It is used extensively by Europeans, in the form of decoctions, for convulsions, lumbago, ringworm, malaria and dysentery. As we have seen, however, great care should be exercised in its use, for the poison solanine is particularly strong in this species.

Altogether there are at least seven species of *Solanum* in the Drakensberg. All Solanums are called um-Tuma by the Zulus.

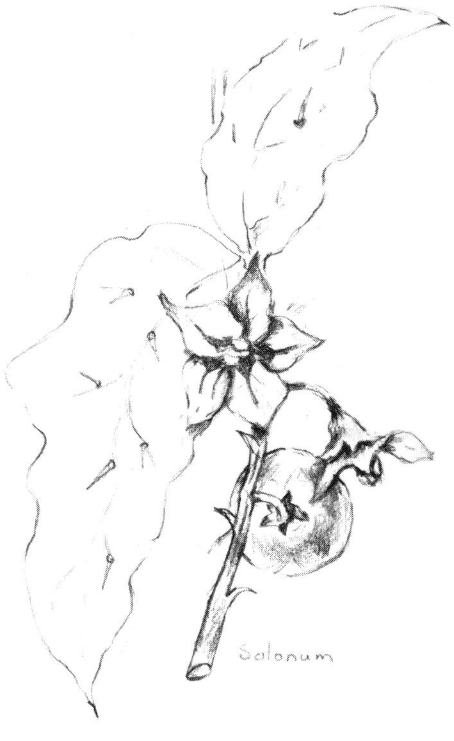

1

2

3

4

1. *Solanum aculeatissimum.*
2. *Solanum nigrum.*
3. Fruit of *Solanum aculeatissimum.*
4. Fruit of *Solanum nigrum.*

SCROPHULARIACEAE

An ugly-sounding name for a group of most lovely plants, for this family includes some of our best-known and most beautiful garden flowers. All our Antirrhinums, Foxgloves, Calceolarias and Pentstemons belong to this family, while digitalin, so useful in treating heart cases, comes from the plant Digitalis, also a member of this family. The name is derived from the Latin word *scrofula,* which means a swelling of the lymph glands. Many of the plants in this family were supposed to be useful in treating this condition. Included also in the family are some shrubs and a few trees. Many of the plants are parasitic or semi-parasitic. Scrophulariaceae is mainly an African family, with a few plants coming from Madagascar and Australia. There are something like 220 genera altogether, with more than 3 000 species.

HALLERIA

Scrophulariaceae

Halleria lucida is one of the tree-like Scrophulariaceae. It grows to a height of five to six metres, and at the end of winter and in the early spring it comes out in dense clusters of bright red or golden-orange tubular flowers. These flower clusters are produced straight from the branches, on both young and old wood, and sometimes even on the main trunk itself, a type of flower production which is called cauliflory. The wood of the tree is fairly hard, and is often used for making kraals, as it is supposed to be impervious to white ants. It was also used by the Zulus for the shafts of their assegais, for the branches are often very straight, and the wood is also supposed to be ideal for producing the friction needed for fire-making. The fruits are bright green and globose, and contain a greenish jelly-like substance in which the seeds are embedded. They later turn a purplish-black before drying out. They are edible.

But it is the flowers themselves which are so striking. They glow in the warm sunlight of early spring like a softly glowing fire, and are much sought after by sunbirds.

The tree grows on the slopes and in the valleys of the Little Berg, often in dongas and ravines and on the forest margins, at an altitude of about 1 250 metres. Common names are Wild Fuchsia, or Tree Fuchsia. The genus itself was named by Linnaeus after Albrecht von Haller, a well-known physician and botanist at Göttingen, in 1736. *Lucida,* in Latin, means bright or dazzling.

The Zulu name is usually given as u-Tshwala-bezinyoni, which means beer for the birds, but Dr E. J. Moll also gives u-Nondomela and u-Nobibi. I have also been given i(li)-Minza and uBinda wehlathi. The Xhosa name is umBinza.

Altogether there are nine species of *Halleria,* found only in Africa and Madagascar. Three come from South Africa, but only *Halleria lucida* occurs in Natal. The genus is widespread in South Africa, being found in all four Provinces, Swaziland, Lesotho and Botswana, and from sea-level to high mountain areas.

The dried leaves of the plant are stored by the Zulus. When required, they are moistened with water and the juice squeezed into the ear for the relief of earache. The Xhosas burn twigs of the tree as a sacrifice to their ancestral spirits.

Halleria lucida. Note, in the two lower pictures, how the flower clusters grow straight from the woody stems and trunk.

DIASCIA *Scrophulariaceae*

Diascia barberae, often cultivated in Natal gardens, is a small, dainty plant, indigenous to South Africa, and is one of the more colourful of our Drakensberg wild flowers. It is usually found growing on the summit plateau in damp places, often in the shade of basalt outcrops. The flowers are a deep rose-pink in colour, and have five petals, the anterior being much larger than the other four, and equipped with two spurs. These two spurs are used in the naming of the plant, for the name Diascia comes from two Greek words, *dis* meaning two, and *askos* which means a small bladder or sack. It grows up to 0,4 m in height, and the flowers are about 15 mm in diameter. It is related to our cultivated Snapdragon.

The word barberae probably comes from Mrs F. W. Barber of Queenstown, who collected the plants actively round about 1860.

Diascias are found mostly in the western Cape, but they also occur in Natal and the Orange Free State. There are none in the Transvaal. In some species the colour merges into purple, but rose-red is the prevailing colour. Altogether there are more than 50 species.

Other Drakensberg species are *D. cordata, D. integerrima, D. stachyoides* and *D. purpurea.*

NEMESIA *Scrophulariaceae*

Nemesia is an African genus, with about 70 species, 62 of which occur in South Africa. It is found throughout the Republic, but its true home is the south-western Cape, where the majority of the species occur. Like *Diascia,* it is very similar to the Snapdragon, so much so that originally some of the species were actually classified under this genus.

The flowers have a short, broad tube which opens into two large 'petals', or more correctly lobes. The upper lobe is divided into four segments, the margin of each segment, or scallop, being curved. The lower lobe is single, but slightly ruffled.

Nemesia coerulea stands about 150 mm high. The flowers are light purple, and, like all Nemesias, are two-lipped, and with a small yellow spur at the base of the perianth segments. The actual flowers are rather like our garden Linarias, but larger, about 15 mm in diameter. The leaves are narrow and serrated.

Nemesia capensis, as the name indicates, is really a Cape species, but it occurs also in the Drakensberg, where it is probably the commonest of all the Nemesias. The flowers are pink, but occasionally blue, lavender and even white varieties are found.

Two other species which occur in the Drakensberg, are *N. melissaefolia* and *N. denticulata.*

A common name for many of the Nemesias is Leeubekkie. The generic name *Nemesia,* is the name used by Dioscorides of Anazarba in his great work on Materia Medica (written about A.D. 77) to describe the Snapdragon.

1

2

3

4

1. *Nemesia coerulea.*
2-4. *Diascia barberae.*

PHYGELIUS

Scrophulariaceae

Phygelius aequalis is a rather showy plant, growing on the banks of streams in the Little Berg, with long stems bearing masses of dusty-pink tubular flowers. The flowers droop down at an angle, pointing inwards, and they bloom from September to March. The common name is River Bells. There is some doubt about the origin of the name Phygelius. It comes either from the Greek word *phyge* meaning flight, because collectors seemed to miss the plant for so long, or *phyge* plus *helios,* the sun.

The Zulus pound the roots, burn them, and apply the ashes to cuts and bruises. The ash is also used in various ways in witchcraft, and in circumcision ceremonies.

The genus is endemic to South Africa. There are two species, *Phygelius aequalis* and *P. capensis*. *P. capensis* is very similar to *P. aequalis,* but the tubular flowers are a little shorter and broader. Its habitat is different, also. *P. aequalis* grows in the valleys of the Little Berg at an altitude of about 1 500 m, while *P. capensis* is usually found at much higher altitudes, having been recorded even from the summit of the main escarpment, at 3 000 m, in the Giant's Castle area, and at Mont-aux-Sources.

Phygelius is essentially a mountain genus, and is fairly common in the Drakensberg, and also in the mountains of the northern Transvaal, Swaziland, Lesotho, the eastern Orange Free State and the Eastern Cape.

The Southern Sotho burn the roots of *Phygelius aequalis,* powder them, and rub the powder into cuts and bruises to relieve the pain. They also use the plant as a charm against hail damage, to protect against witchcraft, and to give courage in war. The Sotho name for the plant is metsi-matso.

SUTERA

Scrophulariaceae

The bright, brick-red 'petals' of *Sutera breviflora* in the green grass of early summer make this a striking flower. It is a small, straggly shrub, only about 150 mm high, bearing deep red flowers. These are small, only about 10 mm in diameter, and are rather like our garden Phlox in shape, with five 'petals' and yellow centres. The leaves are sticky and aromatic. The plant grows on the slopes and on the summit of the Little Berg, at 2 000 metres, and it flowers throughout the summer.

The genus was named in honour of J. R. Suter, Professor of Botany at Berne University. The specific name *breviflora,* comes from the Latin *brevis,* meaning small or short, and *flos (floris)* a flower.

There are over 100 species of *Sutera* altogether, the majority coming from South Africa. One species comes from the Canary Islands. The Drakensberg can lay claim to seven.

1. *Phygelius aequalis.*
2. *Sutera breviflora.*
3. *Phygelius aequalis.*
4. *Sutera breviflora.*
5. Close-up of the flower of *Phygelius aequalis.*

1

3

2

4

5

ZALUZIANSKYA

Scrophulariaceae

In the 16th century there lived in Poland a botanist-physician by the name of Adam Zaluzianskya á Zaluzian (1558-1613). He had no connection whatever with South Africa, as far as I know, and never visited the country, but he gave his name to a genus that is confined to South Africa. This is the genus *Zaluzianskya*. There are 36 species altogether in the genus. Eight of these are known to occur in the Drakensberg, and it is possible that three more (which make up the total of 11 Natal species) may also occur, though I have not actually seen them myself.

Many of the species open only in the late afternoon, some only at night. All plants in the genus have long, slender corolla tubes with five spreading lobes ('petals'). During the day the 'petals' curl inwards to form a rounded ball, and the thin flower-tube looks like the handle of a drumstick. Hence comes the common name for many of these species, Drumsticks. Sima Eliovson remarks on what a pity it is that such a small and dainty flower should have such a long and difficult botanical name. Let us at least be grateful that the flower was not given the gentleman's full name!

Zaluzianskya maritima is, perhaps, the best known of the species. This is a small perennial plant, about 300 mm high, which grows in grasslands at about the 1 800 metre level. It flowers in December and January. The flowers are pinkish-red in colour, with a yellow (sometimes orange) eye. Each flower has a slender tube, opening into a small, flattish face, with five notched and heart-shaped 'petals'. The buds, on long stalks, are bright red and globular. The long-tubed corollas are white and strongly scented at night, thus attracting night-flying moths, which assist in the pollination. (That is why some species open only at night).

Zaluzianskya ovata is a plant that grows amongst the basalt rocks on the summit, at 3 000 metres, usually in damp, shady places. The flowers are white, with red on the undersides of the 'petals', and about 15 mm in diameter. All five 'petals' are deeply notched, and when they curl up, they make the usual rounded red-headed drumstick. The flowers grow together in great masses, sprawling over the basalt rocks in compact groups that are very typical of this particular species.

Zaluzianskya distans has flowers that are very similar to *Z. ovata,* white above and red beneath, but it grows at a lower altitude, usually below the sandstone cliffs of the Little Berg. The plants also grow singly, rather than in clusters, as in the case of *Z. ovata*. The flowers open only in the very late afternoon and at night.* Flowering period is from November to February.

*One interesting point is that this night-time opening is actuated not by the fading of the light, but by some sort of timing mechanism. If the flowers are placed, during the day, inside a dark cupboard, they will still not open until night falls.

SOPUBIA

Scrophulariaceae

On your climbs in the Little Berg, in summer, you will often come across a small, erect plant about 200 mm high, growing in the grasslands. It has silvery-grey foliage (the leaves are thin and long) and small pink flowers, each about 10 mm in diameter, with five petals and four stamens in the centre, yellow in colour. This is *Sopubia cana*. The name Sopubia is probably an anagram for Bopusia, an old name for *Graderia scabra,* another genus in this family, while the specific name comes from the Latin *canus,* meaning greyish-white, referring to the foliage.

Sopubia is chiefly an African genus, with 5 species recorded in South Africa. As far as I know, *Sopubia cana* and *S. simplex* are the only two species to occur in the Drakensberg, the latter having been recorded in the Cathedral Peak area.

1

2

3

4

1. *Zaluzianskya distans.*
2. *Zaluzianskya ovata.*
3. *Sopubia cana.*
4. *Zaluzianskya maritima.*

GRADERIA

Scrophulariaceae

This is a small genus, with only five species, again essentially African, but there is one species recorded from the Island of Socotra. Only *Graderia scabra* occurs in Natal, but elsewhere in South Africa the genus is widespread with three species from the southern and eastern Cape to the Transkei, Swaziland, Transvaal, and on into tropical Africa.

Graderia scabra is a small, dark pink, semi-herbaceous perennial, about 0,3 m high, usually found growing in the grasslands of the Little Berg. Several flower stems rise from a wooded rootstock, and the flowers, which are trumpet-shaped, are borne singly from leaf axils. The plant is semi-parasitic, and it flowers in early spring.

Its common name is Wild Penstemon, and the Zulu name for it is isi-Qomiso. The Zulus make an infusion from the roots and this is poured into the nostrils of young men intent on courting. It is supposed to ensure success and act as a charm. The specific name *scabra* comes from a Latin word meaning rough. The leaves are rough-textured.

CYCNIUM

Scrophulariaceae

The genus *Cycnium* is confined to Central and Southern Africa, and contains over 100 species. Here in South Africa we have five species, and as far as I know only one of these, *Cycnium racemosum,* occurs in the Drakensberg.

It is usually found growing on the banks of mountain streams, between 1 800 and 2 400 metres, and on the grassy slopes below the sandstone cliffs of the Little Berg. The plant is about 0,6 m tall, and each stem bears a cluster of bright pink flowers, each about 30 mm across. When dried, the plant turns black, a tendency which is common in many plants of this family.

Cycnium racemosum is a hemi-parasite (again like so mnay plants in this family), and parasites cannot be cultivated in the absence of their hosts. This is a pity, for this particular plant would grace any South African garden.

The word Cycnium comes from the Latin *cygnus,* a swan. *Racemosum* comes originally from the Latin word *racemus,* a bunch of grapes. From this comes the botanical word racemose, which means a flower-cluster where the separate flowers are attached by short stalks at equal distances along a central stem.

The root is used by the Zulus as an emetic and for the relief of pain. It also has another, rather strange, use. When an animal is to be slaughtered, leaves of this plant are collected and burnt. The animal is then 'doctored' by passing it through the smoke. The Zulu name for the plant is um-Hlabahlangane.

Rather more common than *Cycnium racemosum* is *C. adonense,* the type species, but as far as I know this lovely white species, though widespread, does not occur in the Drakensberg.

---

1 & 2. *Graderia scabra.*
3 & 4. *Cycnium racemosum.*

1

2

4

3

Harveya, the last of the Scrophulariaceae to be illustrated in this book, is an interesting genus. The plants are small, reaching a height of about 0,25 m, and all are parasitic. Like many parasites, they have no green colouring whatever, and no green leaves. The plants obtain their nourishment from the roots of other plants. The petals are thick, wax-like and crenallated (crinkly), and they turn black when touched. The common name of *Harveya speciosa* is Inkblom.

The genus was named after Dr W. H. Harvey (1811-1866), the famous South African botanist who arrived at the Cape in 1835, became Treasurer-General of the Colony, and, in his spare time, commenced the monumental *Flora Capensis,* only completed in 1933, 67 years after his death.

" 'Tis a very lovely plant," said Dr Harvey, when he heard that Sir William Hooker had named the genus after him, "with which I am highly pleased and flattered", adding with a mischievous glint in his eye, " 'Tis *apropos* to give me a genus of parasites, as I am one of those weak characters who draw their pleasures from others, and their support and sustenance too, seeing I quickly pine if I have not someone to torment." "The man from Limerick", adds Conrad Lighton, who tells the story, "who had the will and energy to launch our mammoth *Flora Capensis,* was no parasite!"

We illustrate on the opposite page two species from this genus. *Harveya speciosa* has a cream flower, and occurs on the slopes of the Little Berg, at about the 1 750 metre level. The word speciosa means, in Latin, full of beauty, a fine display, an excellent name for this striking flower. The Zulus pound the roots and mix them with water. The liquid is then used to clean and drain nasal passages. The whole plant is used as a remedy for dizziness and madness, and for the treatment of the bewitched.

Harveya huttonii has flowers of deep pink to scarlet. It grows at somewhat higher altitudes to the last, being usually found on the summit, or in the upper reaches of the passes leading to the summit. Very occasionally it is found along stream banks in the Little Berg. Both species flower in midsummer.

Altogether there are six species of the genus *Harveya* in Natal, 24 in South Africa as a whole. Four of these are found in the Drakensberg, the other two being *H. laxiflora* and *H. pumila*.

Harveya speciosa is used by the Southern Sotho as a remedy for madness, and it is also used by them in treating a person who has been bewitched by his relatives. The Sotho name for the plant is seona, while the Zulus call it umShelezana omhlophe.

1

2

3

1. *Harveya huttonii.*
2. *Harveya speciosa.*
3. *Harveya huttonii.*

SELAGINACEAE

The Selaginaceae family is a small one, consisting of ten genera, only three of which are found in South Africa. These used to be included under Scrophulariaceae, but they are distinguished from the latter by having a single ovule in each of the two chambers of the ovary, while Scrophulariaceae plants have more than one. It was thought better to place these genera in a family of their own. It is essentially an African family, but one or two genera appear in Madagascar. The leaves of most species are small or ericoid. The three South African genera, all of which appear in the Drakensberg, are *Hebenstreitia, Selago* and *Walafrida*.

HEBENSTREITIA *Selaginaceae*

Hebenstreitia is a wholly South African genus. There are eight species in Natal, and it is possible that four of these are to be found in the Drakensberg. They are slender, tapering plants, with a spike of small flowers about 200 mm in length. The corolla is split open along the anterior side and the style and stamens protrude through the slit. The leaves are long and needle-like. Some species are shrubby with woody rootstocks.

The two commonest species in the Drakensberg are *Hebenstreitia fruticosa* and *Hebenstreitia dentata*. The flowers of *H. fruticosa* are white, with an orange throat, clustered in a slender spike. The Zulus call this flower im-Pishi(m)nishi enkulu. They make an infusion from the pounded roots which they sprinkle around their kraals in time of war as a protective measure. Warriors also bathe in the liquid to ward off death in battle. The plant has two common names, Cat's Tail and Slakblom. *Fruticosus,* in Latin, means shrubby or bushy.

Hebenstreitia dentata is similar in shape and form to *H. fruticosa* but the flowers are more yellow in colour. The throat is the same, a deep orange. The flower is said to have no scent in the morning, a strong and disagreeable one at noon and a pleasant one in the evening. *Dentata* means toothed.

Both these species flower in midsummer. They can usually be found in grassland between 1 800 and 2 200 metres, but *H. fruticosa,* at least, has been recorded from the summit.

The genus was named in honour of Professor J. E. Hebenstreit, a Professor at Leipzig University.

SELAGO *Selaginaceae*

Selago flanaganii, commonly known as Blue Haze, is a small blue flower which grows in dense clusters on a 0,5 m high bush on the summit plateau at 3 000 m. The flowers are only a few millimetres in diameter, and consist of a thin tube opening into five starry petals. The general effect, as its common name suggests, is of a misty-blue haze. Although it is essentially a summit species, it is sometimes found growing in grasslands as low as 2 200 metres. It flowers in January and February. Selago is an ancient name for the Club Moss, *Lycopodium,* the small leaves and general appearance of which are very similar to those of *Selago*.

It will be noticed that the flowers in the picture opposite have a pinkish hue. This is an example of the well-known difficulty experienced by photographers in photographing certain blue flowers, with modern colour film. The phenomenon is discussed on page 231.

1. *Selago flanaganii.*
2. *Hebenstreitia fruticosa.*
3. *Hebenstreitia fruticosa.*

1

2

3

GESNERIACEAE

This is the Gloxinia and Streptocarpus family. It is a large one, with something like 700 genera altogether, and 700 to 800 species, widely distributed, especially in tropical and subtropical areas. In South Africa we have only two genera, *Streptocarpus* and *Charadrophila,* the latter, with only one species, and recorded only from a small area, from Jonkershoek near Stellenbosch. *Streptocarpus* has 28 species in South Africa. The flowers are often very showy, and usually tubular or bell-shaped. Many of our greenhouse plants, such as Gloxinia and the African Violet, *Saintpaulia,* belong to this family.

STREPTOCARPUS

Gesneriaceae

Streptocarpus gardenii is a shy, delicate flower that loves the green gloom and the silences of the deep forest. Even in the forests it prefers the south sides of rocks, so as to be always in the shade. The flower has a long tube about 30 mm in length, ending in five petals which flare outwards, giving the flower a somewhat bell-like shape. The flowers are about 25 mm in diameter, and are a pale mauve in colour. The leaves (sometimes a single leaf) are at ground level, and the flowers rise direct from these basal leaves. The anthers face each other in pairs.

The specific name was given in honour of Captain R. J. Garden, who, in 1853, was attached to the 45th Regiment. In that year he was commissioned by the Government to explore the country between the Umzimkulu and Giant's Castle, and he did a certain amount of plant collecting.

The seed-pods of all the plants in this genus are interesting. They have a spiral twist to them, and when this dries out, the spiral relaxes like a spring, and this shoots the seeds out, ensuring that they are scattered. Hence comes the name of the genus, for *strepto,* in Greek, means twisted, and *carpus* means a seed or seed-pod.

The root system is shallow. The plant often grows only on thin moss or on a slight layer of humus on a rock face. Sometimes it even grows on the trunks of forest trees, though it is not, of course, parasitic.

There are five other species of *Streptocarpus* that grow in the Drakensberg, *Streptocarpus pusillus,* which is white in colour, with only one leaf; *Streptocarpus cooperi,* growing only in the Van Reenen area and nowhere else in the Drakensberg; and two fairly rare species, *S. pentherianus* and *S. polyanthus.*

Streptocarpus gardenii, growing in the green gloom of the yellowwood forests.

ACANTHACEAE

The Acanthaceae family includes many of our popular greenhouse and indoor plants. It is a fairly large family, with 180 genera and over 2 000 species. It is abundant in the tropics, but is not found in cold regions. In the whole of Europe there are only a few species of the genus *Acanthus*. (This is the flower the decorative leaves of which figure so prominently in Greek architecture.) In South Africa the family is common in the Transvaal and Natal, but does not reach as far as the Cape Peninsula. The derivation of the family name is from the Greek word *acantho,* which means spiny. Many of the plants in this family have spiny and prickly leaves. In Natal we have 27 genera, of which at least four occur in the Drakensberg.

BARLERIA

Acanthaceae

Most of the Barlerias have bluish-mauve flowers, but there are a few white and yellow species. *Barleria monticola* is a small plant, with mauve flowers, found growing in the lower regions of the Litte Berg flowering often after the veld has been burnt. The flower itself is about 25 mm in diameter, with a short tube flaring out into five petals, and the plant itself is from 100 to 150 mm high.

Barlerias are widely distributed throughout Natal, Asia and America. Here in Natal we have 14 species, but as far as I know only one of these, *Barleria monticola,* is found in the Drakensberg.

The genus was named in honour of Jacques Barrelier (1606-1673), a well-known French botanist. *Monticola* means growing in mountains.

An infusion is made from the roots by the Zulus and used by young men when courting to *phalaza* with, but this must be done just when the sun is rising.

Barleria monticola

RUBIACEAE

This is the Gardenia family, with a wide distribution in all parts of the world, especially in the tropics. Altogether there are over 400 genera, with from 5 000 to 6 000 species. Natal has 45 genera and 132 species. Nine of these genera are represented in the Drakensberg. Many beautiful garden plants belong to this family, and commercially we have the coffee plant, *Coffea arabica,* and *Cinchona,* several species of which produce the quinine so useful in combating malaria.

PENTANISIA

Rubiaceae

Pentanisia is one of the commonest flowers in the Drakensberg, certainly in the Little Berg, where it is widespread. In South Africa there are only two species of the genus, *Pentanisia prunelloides* and *P. angustifolia,* the former being the better known of the two. Actually there is little difference between them. *Pentanisia prunelloides* is simply a rather darker blue than *P. angustifolia.*

Pentanisia prunelloides is a small, tufted plant, with a rounded head of bluish-mauve flowers at the end of each stem. The stems vary in length from a few centimetres to 0,3 m, and the flower heads are about 40 mm in diameter. The leaves are slightly velvety.

Common names are Wild Verbena and Sooibrandbossie. The Zulu name is i-Cimamlilo, which means put out the fire.

The specific name *prunelloides* (meaning resembling Prunella, a genus of the Labiatae) probably comes from a Latin word meaning a small plum or prune, but there is a rather strange coincidence relating to this specific name. There is another Latin word, *pruna,* which means pertaining to fires. Compare this with the Afrikaans name for the plant, Sooibrandbossie, and the Zulu name, i-Cimamlilo. The derivation of the name Pentanisia is also a little doubtful. Could it possibly come from the Latin word *penta,* meaning five, and referring to the five perianth segments, and Nisus, king of Megara, a gentleman who sported bright purple hair?

The plant is much used medicinally. The Zulus make an infusion from the roots and use it as an enema for stomach pains. They make a poultice from the leaves for the relief of painful swellings, rheumatic pains and sprains. Europeans, too, have used the plant with considerable success, I believe, in the treatment of rheumatoid arthritis. Bryant records that the Zulus also use the roots as a remedy for piles. In the Transvaal Africans use the plant as a bath-salt. A decoction of the plant is added to the bath to enhance its refreshing qualities. The Sotho use it as a charm. Mixed with other ingredients it is used to prevent a sorcerer from finding the door of the hut.

Pentanisia prunelloides

As you make your way through the forests of the Little Berg, especially along stream banks, you will sometimes come across a small, rounded tree, about four to five metres high, with dark green shining leaves, and bright corymbs of deep scarlet or orange tubular flowers. The bark is a little pale and the wood hard and heavy. This is *Burchellia bubalina* (formerly *B. capensis*).

Burchellia bubalina is the only species in the genus, and it is widespread along the southern and eastern regions of South Africa, and reaching well up into the mountain regions. It is endemic to South Africa. September is the flowering time.

Each flower contains a sweet, heavy nectar, much loved by the birds, and hence comes the Zulu name u-Tshwala-bezinyoni, which means beer for the birds. Two other Drakensberg flowers have the same name, *Halleria lucida* and *Leonotis leonurus*. To distinguish between *Burchellia bubalina* and the Leonotis flower, the Zulus add the word 'obukhulu' (large) after u-Tshwala-bezinyoni in the case of *Burchellia bubalina,* and 'obncane' (the small one) in the case of *Leonotis*. A cold water infusion is made from the pounded roots by the Zulus and this is supposed to act as a cure for bad dreams. They also use the root as a nasal and body wash, as an emetic, and also for love philtres.

Common names for the tree are Wild Pomegranate, Buffelshoring and Buffelshoorn. The genetic name honours William J. Burchell (1782-1863), the famous South African traveller and explorer in the early 19th century, who is reputed to have collected over 40 000 plants during his travels. The specific name *bubalina* is something of a puzzle. It comes from the Latin word *bubulus* which means pertaining to oxen (buffalo), but what the connection is, is not clear. Some claim that the fruit vessels have pointed lobes resembling the horns of miniature buffaloes, but this is doubtful. In at least one other case *(Euphorbia bubalina)* the word *bubalina* describes the habitat of the plant, near the Buffalo River, and this could apply in the case of *Burchellia bubalina,* for it is found along the coast of the south-east Cape. But there is yet another derivation. I have been told that Burchell set out one day to collect plants on a hill, at the top of which was a small forest. As he approached the forest he saw what he took to be a dark buffalo in the shade of marginal trees. When he got nearer he found it was a dark-leaved shrub, which he collected. When John Sims, the Editor of the Curtis Botanical Magazine, came to name the plant, he called it *Burchellia bubalina,* or Burchell's Buffalo. It is a good story, but whether it is true I cannot say.

Burchellia bubalina

PAVETTA *Rubiaceae*

Pavetta cooperi is an ever-green shrub or small tree, two or three metres in height, and, like *Burchellia bubalina,* is often found growing on the forest margins. It has rounded heads of pure white flowers, with long delicate white styles protruding from each flower, giving it a spiky appearance. (This is typical of all the species in this genus). The flowers are sweetly scented and appear at Christmas time, hence the common name Christmas Bush. Another common name is Kaffir Bride. The leaves are dark glossy green, oval and tapering. They also have small raised bacterial nodules on the surfaces. These bacteria are harmless to the plant, and it is thought that they actually assist the plant to assimilate the free nitrogen in the air.

The genus has over 350 species, about 35 of which are indigenous to South Africa. They grow in all four Provinces, but not in the south-western Cape. Natal is home to 25 species, but as far as I am aware only one *Pavetta cooperi,* occurs in the Drakensberg, though *P. assimilis* and *P. lanceolata* may occur. Most of the species in Natal are found along the coast and in Zululand.

The Zulu name is is-Anywana. An infusion is made from the roots of this plant, together with roots from another plant, umanzamnyama, *(Anemone fanninii).* If the members of one kraal have a grudge against another kraal they *phalaza* with this mixture just before a beer drink. This is supposed to cause the deaths of the children of those they hate.

The origin of the name Pavetta is obscure. It may be the vernacular name of a plant in Malabar. There is a Latin word *pavito* which means to tremble or quake with fear, but this would appear to have no connection with the plant.

CONOSTOMIUM *Rubiaceae*

Conostomium natalense var. glabrum is a small shrubby plant, about 0,3 m high, with small bluish flowers about 10 mm in diameter. The flowers form a cluster at the tip of a short stem. They have four petals (very occasionally five) with dark brown and cream centres. The petals are oval and pointed. The leaves are also pointed, and narrow, broader at the base and about 40 mm long. There is a somewhat rarer, white variety, and the flowers come out in late summer.

This species used to be placed in *Oldenlandia,* the genus named after Hendrich B. Oldenland, a Dane in charge of the Dutch East India Company's garden in Cape Town during the time of Simon van der Stel, but a few years ago it was taken out of this genus and placed in the genus *Conostomium* which now has nine species. Two of these occur in South Africa.

Pavetta

1

2

3

1. *Conostomium natalense var. glabrum,* the white variety.
2. *Conostomium natalense var. glabrum,* the pale mauve variety.
3. *Pavetta cooperi.*

DIPSACACEAE

This is the Scabious family, a family of 9 genera and more than 300 species. It is abundant in the Mediterranean region, but is also found in Africa, Asia and Europe. South Africa has only two genera. The inflorescence is always a compact head, surrounded by an involucre of bracts. The name comes from the Greek word, *dipsa,* meaning thirst but what the connection is, is not very clear.

SCABIOSA AND CEPHALARIA

Dipsacaceae

The two South African genera, both represented in the Drakensberg, are *Cephalaria* and *Scabiosa*. They are very closely allied and at first glance are difficult to differentiate. The flowers in both genera form compact heads of tiny flowerets on a disc, each head being about 25 mm across, and almost always white, though *Scabiosa* can sometimes be pink or mauve and there are a few blue species of *Cephalaria*. In *Cephalaria* the calyx is small and pappus-like, with short scales, and there are several rows of involucral bracts. In *Scabiosa* the calyx consists of five long bristle-like awns and only one or two rows of involucral bracts. These awns project up between the softer flower parts. They are particularly noticeable when the flowers are in bud or after the corollas have fallen. The leaves of *Scabiosa,* in general, are much narrower than in *Cephalaria,* the plant is branched, and the flower-heads stand up well above the foliage. The whole plant is about 0,5 m in height. In *Cephalaria* the leaves are much broader, toothed, and they form a rosette at the base of the plant. Like *Scabiosa* the flower-heads are borne on long, slender stalks, about 0,3 m in height.

The *Cephalaria* illustrated opposite is *Cephalaria oblongifolia* (it used to be called *Cephalaria attenuata*). It is usually found on the slopes of the Little Berg during January and February. The word Cephalaria comes from a Greek word *kephale,* meaning a head. There is one other Drakensberg species that I know of, *Cephalaria natalensis*. Both species flower during the latter part of summer. In some areas the Zulus pound the leaves, squeeze out the juice, and a few drops are given to a child suffering from a bad cold.

Scabiosa columbaria is well-known in South Africa, appearing all the way from the Cape Peninsula to the northern Transvaal and on into Ethiopia. The leaves are deeply lobed and toothed (this is a distinguishing feature of the plant) and it grows at a slightly higher altitude to the other Drakensberg species, *Scabiosa drakensbergensis,* and the Cephalarias – usually on basalt outcrops at about 1 800 to 2 700 metre levels. It flowers during February and March. *Scabiosa drakensbergensis* is often found on stream banks and moist gullies in the Little Berg. The word Scabiosa comes from the Latin word *scabiosus,* meaning scabby or scurfy. Some European species of the plant were supposed to cure scabies, or the itch.

The root of *Scabiosa columbaria* is used medicinally by many Africans. It is either chewed, or used as a decoction, as a remedy for colic. This has been tested, and confirmed, by several European medical practitioners. The Sothos use a decoction from the root as a remedy for sterility, and they also make an ointment from the charred root for the treatment of venereal sores. The Xhosas use a preparation from the root for sore eyes, while the dried plant is often used by Africans as a perfumed dusting powder, especially for infants. The powder is known as *makgha*. The Xhosa name is isiLawu esinkhulu (sometimes esimhlope).

Common names are Bitterbos, Rice Flower, Morning Bride, Wild Scabious and Pisgoedbossie. You will see the significance of the latter name if you remember that the plant can be used for urinary disorders!

1

2

3

1 & 2. *Scabiosa columbaria*.
3. *Cephalaria oblongifolia*.

CAMPANULACEAE

Campanulaceae, the Blue-bell family, is widespread throughout the world. Altogether there are something like 70 genera, with over 1 500 species. South Africa can lay claim to 16 of these, with seven coming from Natal. They are most prolific, however, in the Cape. The flowers are usually blue in colour, some, though not all, bell-shaped *(campana* is the Latin word for bell), and many of the plants have latex. Our well-known garden plants, the Canterbury Bell, the Blue Bell of Scotland, and our Lobelias are all members of this family.

LOBELIA

Campanulaceae

One cannot fail to notice the bright blue eyes of the dainty little *Lobelia,* starring the grasslands of the Little Berg in early summer. Any leisurely climb in the lower foothills will reveal masses of them. A low-growing plant, one often finds them growing in great sheets of colour, usually in damp hollows and wet places.

The genus *Lobelia* is a large one. Altogether there are up to 300 species, with about 75 coming from South Africa. At least six are found in the Drakensberg, of which we illustrate three.

The genus was named after Mathias de l'Obel (1538-1616), a physician to William the Silent and later botanist to James I of England. Lobelias range in size from tiny plants only a few centimetres high to giant ones in tropical Africa, three metres and more in height. Each flower has a short tube, a two-lobed upper lip and a three-lobed lower lip. This makes it a very easy flower to identify.

Lobelia decipiens, the Butterfly Lobelia, is a small, dainty plant, growing in moist places in the Little Berg. It is about 150 to 200 mm high, and is easily recognised from its colour. The three-lobed lower lip is a dark blue, the two-lobed upper lip a rich purple, and there is a yellow spot in the throat. The stems, mostly single, arise from a creeping rhizome. The word decipiens comes from the Latin word *decipio,* which means to catch, deceive or ensnare, but what connection this has with the plant I have no idea.

The Zulus, who call the plant isi-Dala ikhambi lomkhuhlane, make an infusion from the whole plant and this is taken in small doses for colds. It is also believed to be a cure for skin diseases, gout and rheumatism.

Lobelia filiformis var. natalensis has pale blue flowers, very similar in shape and size to *Lobelia decipiens,* each flower with a fairly large white centre, and four dark spots near the edge of the white centre. The plant is from 150 to 250 mm high, and flowers during February and March.

Lobelia vanreenensis is a small, delicate plant, found growing at about the 1 600 metre level, in damp places and along stream banks, often in the shade of forests. The flowers are white, with a small yellow centre, and they usually appear in early autumn.

1

2

3

4

5

1 & 2. Lobelia vanreenensis.
3. Lobelia decipiens.
4. Lobelia filiformis var. natalensis.
5. Lobelia decipiens.

The genus *Wahlenbergia* is easily recognised, for the flowers are almost always blue, and they are shaped like the English Blue-Bells. There are five petals, broad at the base and pointed, and these curve outwards, giving the flower its bell-like appearance. A few white flowers are sometimes found. The genus is widespread throughout the world, and in the Drakensberg we have at least ten species. These are sometimes a little difficult to differentiate, for they often merge into each other.

Wahlenbergia undulata is one of the commonest species in our Natal Drakensberg. It is a small plant, about 0,3 to 0,45 m high. The stems are slender and wiry, and the flowers, about six to a plant, are borne at the tips of these slender stalks. It grows at somewhat higher altitudes to the other Drakensberg species, from the summit of the Little Berg up to 3 000 metres. It is usually found in fairly damp places, and flowers during January and February.

The Zulu name is u-Qhaboza. An infusion is made from the pounded roots, and this is used to wash the body with before going into battle. It is supposed to act as a protective charm. It is also used sometimes as an eye lotion. The Southern Sotho use the plant as a spinach.

Wahlenbergia grandiflora has the usual small pale blue, bell-shaped flowers and slender stems. The leaves are small, and borne at intervals up the lower portion of the stem. The Zulu name is um-Nqantula. The sinew, or bridle under the tongue is called, by the Zulus, the umnqantula. If a young pup is ailing, or refusing its food, the Zulu will often remove this sinew, considering this a perfect cure. The pup is then given an infusion made from the roots of this plant, hence the Zulu name for it. There is a white variety of this flower known to the Zulus as u-Shayindoda omhlophe. Herdboys who have let the cattle stray into the lands, chew the roots of this plant to make them, as they believe, invisible to their wrathful fathers.

Wahlenbergia zeyheri has flowers that are of a darker blue than the others mentioned here. In the Drakensberg it is only found above the 2 400 metre level, but in northern Zululand it is found near the coast. It flowers in February. The specific name honours Carl Zeyher, a well-known German plant collector in the Cape round about 1822. Its Zulu name is u-Qhaboza obncane.

The genus was named after Professor Georg Wahlenberg (b. 1784), Professor of Botany at Uppsala, Sweden.

Wahlenbergia undulata

COMPOSITAE

This is the Daisy family, one of the largest of all our natural orders. It is distributed world-wide and is regarded by botanists as the most highly evolved and successful family of all the flowering plants. Altogether there are something like 900 genera, with 17 000 species. In Natal with over 120 genera and 600 species, members of this family comprise roughly 12% of all Natal flowering plants. Compositae 'flowers' actually consist of a mass of small, compacted flowers aggregated into heads, called the disc flowers, or florets (the 'centre' of the 'flower') and an outer whorl of spreading ray flowers, called, wrongly, the petals. There are other families in South Africa with a similar flower structure, e.g. Proteaceae, but these all have *superior* ovaries. The disc florets in the Compositae family have *inferior* ovaries. Widespread as it is, there are very few plants in this family with any commercial value, though we might note the sunflower, with its valuable oil, and the lettuce, *Lactuca sativa,* and, of course, so many of our lovely garden flowers, such as Aster, Dahlia, Gazania and Cineraria.

CALLILEPIS

Compositae

One of the loveliest sights I know is a green hillside in the Drakensberg, on an early spring morning, with the hillside dotted with those shining white flowers which you probably know as Ox-eye Daisies, or Black-eyed Susans and which the botanist calls *Callilepis laureola.* They have creamy-white 'petals' (the ray florets mentioned above) and dark purplish-black centres (the disc-florets). These flower-heads, about 40 mm in diameter, are borne at the tips of long slender stems. The plant has a large, obconical, woody rootstock, thick and knotted, about 0,3 m deep.

It has two Zulu names, i-Hlamvu (this, apparently, is a general name given by the Zulus to any plant that is used to heal sterility), and im-Pila, meaning good health (but see below!) An infusion from the roots is given to barren women to aid conception. But actually the root contains a deadly poison, which has so far defied analysis. We have referred in the Introduction to the possible danger of using indigenous plants with alleged curative powers. In the case of this particular plant the fact that an unidentified poison was at work was first recognised by Bryant in 1909, and then, about 20 years ago, by Dr J. Wainwright, Professor of Pathology at Natal University's Medical School. A number of unexplained Zulu fatalities began to occur, necrosis of the liver being one of the principal symptoms. King Edward VIII Hospital in Durban found themselves treating from 30 to 40 cases a year. Dr Wainwright was joined by Dr Schonland and Mr H. A. Candy, and a detailed investigation was commenced. Close questioning of survivors began to point the finger of suspicion at *Callilepis laureola,* and this was soon confirmed. An overdose of the infusion causes convulsions and coma and a drop in blood-sugar level, followed by liver and kidney failure. It is possible that when the chemical is thoroughly understood it may prove to have a whole range of beneficial therapeutic values, but in the meantime the plant should never be used medicinally in any way. Sinister as it is, the flowers are a lovely sight in the early spring, nodding their dewladen 'petals' in the morning breeze.

The genus consists of only three species, all of which occur in the Transvaal. Only *Callilepis laureola* occurs in Natal. (This is the generally accepted view, but *Callilepis leptophylla has* been collected twice in Natal, once in the Oribi Gorge and once on the banks of the Tugela River below Colenso.)

Calli is the Greek name for beautiful, and *laureola* means a small crown or garland.

Callilepis laureola

The crisp, sunny days of autumn are heralded on the Highveld, along the roadsides of Natal and in the Drakensberg by the glorious trumpet-call of the wild Cosmos. It is as if summer itself were making a last triumphant outpouring of colour before the cold winds of winter set in, and the fields and the roadsides glow richly with the pinks, the whites and the reds of the Cosmos. Sometimes the sea of colour stretches as far as the eye can reach. I have seen acres and acres of it in the Singati Valley.

The botanical name for this lovely flower is *Bidens formosa*. Bidens is a Latin word meaning with two teeth, and this, almost certainly, refers to the two or more stiff-barbed bristles which comprise the pappus, and which cause the fruit to cling to the coats of animals (Cf. the Black-jack, *Bidens pilosa). Formosa,* in Latin, means beautiful or handsome. The flowers are always either pink, white or a deep wine-red, but there is also, I believe, a very rare yellow variety.

Actually the Cosmos is not indigenous to South Africa. How it reached here is something of a mystery. It is thought that it entered the country from West India or Central America in some fodder for the horses of the British cavalry during the Anglo-Boer War. It has spread to Natal from the Transvaal, mostly along the main railway line. It was first reported in 1904, near Pretoria and along the railway lines near Roodepoort, Florida and Zuurfontein. From here it spread to Natal, and is rapidly reaching the coast. In 1973 a few plants were seen for the first time at Inchanga on the main Durban-Pietermaritzburg road.

The young plants of *Bidens formosa* are relished by cattle, and they are said to have a high food value. They are rich in a volatile oil, however, which gives the plant a somewhat unpleasant odour and which may taint the milk. The leaf has a bitter taste. In West Africa the warmed juice of the fresh plant is used as drops for earache and for conjunctivitis, and also as a styptic on wounds.

We have already seen that a close relative to the Cosmos, *Bidens pilosa,* is the well-known and detested Black-jack weed. This was probably brought to South Africa by Africans from the old Portuguese settlement at Delagoa Bay, which, in turn, must have received it from America, where it is indigenous. The name Cosmos comes from the Greek word *kosmos,* meaning an ornament.

1

2

3

Harbinger of autumn, the Cosmos, *Bidens formosa,* splashes the land with a riot of colour. No. 5 was taken in the Cathkin area, with Cathkin Peak in the background, and No. 3 in the Singati Valley.

4

5

Vernonia hirsuta is sure to catch your eye in the Drakensberg, for it looks like the upper half of a large brilliantly-purple puffy ball. It stands about 0,6 m high, and it flowers from October to early December. It is often found growing in small gullies and rocky outcrops just below the sandstone cliffs. The leaves are crinkled, with undulating margins, and have silky hairs on the under surfaces. They are large and tapering. The common name for the flower is Wild Heliotrope.

The Zulus dry out the stalks and leaves and then grind them into a powder. This is used to snuff up the nostrils as a cure for headaches. They call the plant u(lu)-Hlunguhlungwana lwentaba.

The genus was named after a North American botanist, William Vernon. *Hirsuta* means in Latin hairy. The genus is widespread throughout South Africa, except in the south-western Cape. Five species have been identified in the Drakensberg. The Southern Sotho name is hlelehlele sethsee.

ATHANASIA

Compositae

One of the loveliest of our autumn flowers is *Athanasia calva*. Commonly known as the Coulter-bush from the way the plants pack up against the coulter of the plough, it will be found blooming round about April. It is a fairly large bush, about a metre high, with ericoid leaves. The flowers form fluffy yellow balls, mostly in corymbs. The leaves are aromatic when crushed, and are conspicuously gland-dotted. The flowers have a strong honey scent. *Athanasia calva* used to be called *Athanasia punctata*.

I know of three other species of *Athanasia* that occur in the Drakensberg. *Athanasia thodei* (commemorating Hans Justus Thode: see page 152) is a high altitude species, growing amongst the basalt rocks between 2 350 and 3 170 m. It flowers in March and April. *A. leucoclada (leucoclada* means with white stems) is more a low-level species, but it is found in the Drakensberg up to about 1 675 m. *A. tridens* has been reported from both the Giant's Castle and Ndedema areas. It is usually found between the 1 525 and 2 435 m levels.

Athanos is a Greek word, and means immortal, while *punctata* in Latin means punctured. *Calva* is the Latin word for bald.

I have been unable to discover any medicinal use for *Athanasia calva,* but the Zulus in the Mfongosi district use another species, *Athanasia acerosa* to ward off lightning. A portion of the plant is first burnt on a piece of broken pot either inside the hut or just above the kraal, and the ashes are then sprinkled in the yard.

The rich colour of these brilliant yellow flowers makes a memorable sight as the last days of summer give way to autumn.

---

1. *Vernonia hirsuta.*
2. *Athanasia calva* growing beside *Leonotis leonurus.*
3. *Vernonia hirsuta.*
4 & 5. *Athanasia calva.*

Helichrysum is a large genus in this, one of the largest families of the flowering plants, the Compositae. In South Africa we have well over 200 species occurring in the genus, possibly up to 250, while Natal can claim 123 of them. Possibly up to 90 of these occur in the Drakensberg. Here it is only possible to give a very brief, but we hope representative, selection of the many lovely flowers in this genus.

Helichrysums are, generally speaking, well adapted to resist drought conditions, and are usually found growing on rocky, well-drained slopes. They occur in all three of the Drakensberg vegetation zones, but are particularly common on the slopes of the Little Berg and on the summit of the Little Berg. They range from small, low-growing herbs to large woody shrubs, up to a metre in height.

The word Helichrysum means sun-gold, and comes from two Greek words, *helios,* the sun and *chrysos,* gold, an allusion to the bright yellow colours of so many of the flowers in this genus. Practically all Helichrysums are used medicinally by the Zulus and the Xhosas, often in the strangest ways.

Most people associate the Helichrysums with our Everlastings, and they are correct, but there are many other plants included in the genus which are not at all like Everlastings. Amongst the best-known of the Everlastings is *Helichrysum adenocarpum.* This is a bright scarlet flower, but there is a white variety. A characteristic of this plant is the way the many stems branch out in a curved, bell-shaped form from a single point. It is confined to mountain areas mainly, but does occur at lower levels. It is about 200 mm high, and the leaves are whitish and woolly. The Zulu name is u-Hlambahloshane omncane obomvu, while the botanical name comes from two Greek words, *adene,* a gland and *karpos,* fruit or seed.

Another scarlet *Helichrysum* is *H. vernum.* Here the flower-heads are somewhat larger than in *H. adenocarpum,* but the plant is smaller, only about 50 mm high. In *H. vernum* the perianth segments (actually ray florets) are spread out, while in *H. adenocarpum* they tend to turn inwards.

Helichrysum aureum var. scopulosum (the old *H. scopulosum)* is a bright yellow *Helichrysum,* which usually grows on the summit of the Little Berg. Like *H. vernum* it is a small plant, only a few centimetres high, and it usually flowers from July to December.

Quite a number of the Everlastings are white, perhaps the best-known being *H. confertifolium.* This species is found growing in the valleys of the Little Berg. It may be found flowering during February and March.

1

2

3

4

1. *Helichrysum vernum.*
2. *Helichrysum adenocarpum.*
3. *Helichrysum confertifolium.*
4. *Helichrysum aureum var. scopulosum.*

No-one who has ever climbed above the 2 250 metre level in the Drakensberg can fail to have noticed the white *Helichrysum sutherlandii.* The basalt rock faces at these levels are festooned, in the latter half of the summer, with clinging masses of this lovely flower. The actual flower-heads are small, only about 6 or 7 mm in diameter, with yellow centres, but they are closely massed, and together with the bluish-grey foliage against the grey of the rocks they make a striking sight. They begin to flower about the middle of February, are at their best in April, and fade in July.

Another common flower, but in the lower foothills of the Drakensberg this time, is *Helichrysum cooperi.* It flowers in early spring, and usually grows in damp places, especially where the soil has been disturbed. It is a robust, bushy plant, the branches terminating in masses of bright yellow, everlasting-type flower-heads. Leaves and stems are glandular, and slightly hairy, with a strong aromatic scent. A typical feature of the plant is the masses of dead leaves which clothe the lower portions of the stems. It is a common plant in the Little Berg, and is also found in the Free State and Swaziland, and along the Natal coast.

Helichrysum odoratissimum is another yellow Helichrysum, again common on the slopes of the Little Berg in summer. It is a small plant, shrubby in character, sometimes growing singly but sometimes in clumps, with small heads of yellow flowers. The leaves are greyish-green in colour. The specific name, *odoratissimum,* means in Latin very fragrant, and comes from the Latin word *odoratus,* meaning sweet-smelling, fragrant.

The summit of the main escarpment is well above the treeline, and if it were not for *Helichrysum trilineatum* the mountaineer would be hard put to find fuel for his cooking. This is a small, shrubby plant, very common on the summit, growing amongst the rocks, with small yellow flower-heads. It has tough, wiry branches, not very long (about 300 mm) but often up to 25 mm in diameter. These branches have a high oil and resin content, and especially when dried make an excellent fuel. Actually there are two varieties of this plant. *H. trilineatum* grows just below the final summit cliffs. On the actual summit another variety grows, slightly different to *H. trilineatum,* once called *H. trilineatum var. tomentosum,* but this name has now been discarded.

H. trilineatum is often confused with *H. tenuifolium,* but they flower at different times. *H. trilineatum* flowers in early spring, from August to October, while *H. tenuifolium* does not flower until mid-November or early December.

Other species of *Helichrysum* that we might notice are *H. herbaceum, H. appendiculatum, H. marginatum, H. tenax* and *H. acutatum. H. herbaceum* is a golden-yellow Everlasting, about 150 mm high. In the Drakensberg it flowers from December to February, but in the Eastern Province, near Humansdorp, it flowers in June and July, and makes a brave show, growing along the sides of the roads.

H. appendiculatum is a plant with small white flower-heads, in large clusters, about 200 mm high. It flowers in February, and is usually found at about the 1 750 m level. The specific name comes from the Latin word *appendicula,* meaning a small appendage.

H. marginatum is another red Everlasting. It flowers throughout September and October, usually on the summit of the Little Berg. *Marginatum* is the Latin word for broad-brimmed.

H. tenax is a fairly large bush, about a metre high, with yellow flower-heads and sticky leaves. It flowers from September to November, in the lower reaches of the Little Berg, at about the 1 800 m level. The picture on page 214, no. 2, is of a particularly fine field of these flowers I came across one November, many years ago, in the Cathedral area.

H. acutatum is particularly handsome, growing on the slopes of the Little Berg, and flowering in December. The flowers grow in clumps, about 0,4 m high, with massed yellow flower-heads, about 40 mm in diameter.

Helichrysum cooperi

1

2

4

3

5

6

7

8

9

1. *Helichrysum sutherlandii.*
2. *Helichrysum tenax.*
3. *Helichrysum trilineatum.*
4. *Helichrysum odorattissimum.*
5. *Helichrysum sutherlandii.*

6 & 7. *Helichrysum acutatum.*

8. *Helichrysum appendiculatum.*
9. *Helichrysum squamosum.*

GAZANIA

Compositae

The rich golden colours of *Gazania* make a memorable sight in the Drakensberg whenever you are fortunate enough to come across them, for although they are common in the rest of South Africa, especially in Namaqualand, they are somewhat rare in the Drakensberg. The genus *Gazania* used to be a large one, but the more than 40 species that used to be listed as occurring in South Africa have now been reduced to 16, and only three of these occur in Natal, two of them in the Drakensberg. *Gazania* is, of course, a well-known garden plant, low-growing, with masses of rich yellow, golden or red flower-heads. These are large, up to 50 mm in diameter, and the ray florets are often spotted at the base, making a beautiful border round the disc florets. The plants hybridize easily even in nature.

There is a slight doubt as to the derivation of the generic name *Gazania*. The genus was probably named after Theodore Gaza, but as the word *gaza* in Greek means riches, the name is doubly appropriate.

Probably the best known of the Gazanias is *Gazania krebsiana subsp. serrulata,* the Terracotta Gazania, the only South African species to occur also outside South Africa. It usually has bright orange flower-heads with orange centres, but the colour is variable. In Namaqualand it is usually red, while even a white variety is known. The Drakensberg species appears to be always a rich orange, with a dark ring round the central disc florets. *Gazania krebsiana* is now really a collective name, covering the old *G. pavonia, G. pinnata* and *G. longiscapa,* all of which have now been merged into *Gazania krebsiana*.

The species was named in honour of George L. E. Krebs (1792-1844), who emigrated from Germany to South Africa in 1819 and settled in the Bedford district. He became a noted ornithologist and plant collector. (See page 158).

Gazania linearis, the other Drakensberg species, the Yellow Gazania, has long narrow leaves, dark green above and paler below, slightly hairy. Both leaves and flowers spring from ground level, and the whole flower-head is yellow, with no dark markings. The Xhosas peel off the white under-surfaces of the leaves from which they make a thread for their beads.

The Zulu name for both plants is ubu-Bendle. An infusion is made from the roots and used by young Zulu men to *phalaza* with when courting.

The third Natal species, *G. rigens,* is a coast plant.

BERKHEYA

Compositae

Berkheya flowers are easily recognised, for most have yellow, dandelion-type flower-heads and spiny, prickly leaves rather like those of a thistle. They are quite common, and are widespread throughout South Africa. The genus is almost entirely African. There are no less than 31 species in Natal, with numerous varieties and sub-species. At least ten are known to occur in the Drakensberg. We illustrate two species here, *Berkheya setifera* and *B. onopordifolia*.

The name honours Jan de Franq van Berkhey, a Dutch botanist of the 18th century.

HYPOCHOERIS

Compositae

Also illustrated opposite is *Hypochoeris radicata,* an introduced weed. We include it in this section only because it is so common in the Drakensberg, and sure to be noticed by visitors. It is a dandelion-type of flower, with the flower-heads borne on long, slender stalks, about 250 mm long, with a rosette of leaves at ground level.

---

1. *Berkheya setifera.*
2. *Berkheya onopordifolia.*
3. *Gazania krebsiana subsp. serrulata.*
4. *Gazania linearis.*
5. *Hypochoeris radicata.*

6 & 7. *Gazania krebsiana subsp. serrulata.*

1

2

3

4

5

6

7

When autumn comes to the forests of the Drakensberg you will not fail to note the forest trees festooned with great clusters of a bright yellow creeper, the flowers making splashes of brilliant colour against the dark green of the forest. This is the Canary Creeper, *Senecio tamoides*. The 'flowers' are massed into broad compact heads, measuring 120 mm across. The leaves are bright green and glabrous, the stems smooth and somewhat fleshy, and the flowers daisy-like. It has become a popular creeper in many of our Natal gardens. The Zulus call it u-Boqoboqwana olukhulu, and they use the leaves medicinally. These are pounded up and steeped in cold water, to which a little salt has been added. The liquid is then strained through folded grass and used as an enema for flatulence.

Senecio is one of the largest genera in the family of the Compositae. No less than 124 species are listed as coming from Natal alone, with over 2 000 species spread throughout the world. It is also a very confusing genus. Not only is there a very wide range of flowers in it, but there is a good deal of disagreement amongst botanists as to the various species. Name changes have been frequent, and many plants which in the past were classified under *Senecio* have now been transferred to other genera.

The seeds of all Senecios are accompanied by a white down, or pappus, like the hairs of an old man, and from this comes the name of the genus, for *senex*, in Latin, means an old man. These fluffy hairs ('thistle-down') are used by the plant for the dispersal of its seed by wind. Many of the Senecios are poisonous, producing liver damage in animals and man. Great care has to be taken in harvesting wheat to ensure that no *Senecio* seeds are incorporated. *Senecio* flowers are almost always yellow, but there are a few mauve and purple ones.

One of the finest of the purple 'flowers' is *Senecio speciosus*, illustrated opposite. Although occurring sparsely, it is widespread throughout the Drakensberg. I have found specimens growing on the summit plateau at over 3 000 metres, as well as in the valleys of the Little Berg, at 1 250 metres. The plant is about 200 mm high, and likes damp places. It flowers in midsummer.

The Zulu name for *Senecio speciosus* is i(li)-Gwayi. (uGwayi is the Zulu word for tobacco or snuff), and the Zulus use the leaves for making a snuff which is supposed to relieve headaches. The specific botanical name comes from the Latin, and means beautiful, or showy.

1. *Senecio speciosus.*
2. *Senecio tamoides.*
3. *Senecio* sp.
4. *Senecio tamoides.*
5. *Senecio speciosus.*

1

2

3

4

5

One of the commonest Senecios in the Drakensberg is *Senecio bupleuroides*. It is especially prolific where ground has been disturbed, and in the spring the massed plants cover the low-lying lands of the Little Berg in a lemon-yellow sheen. The 'flowers' are yellow and daisy-like, and borne on a single stalk, nearly a metre high, which branches at the top into a number of stems, each about 250 mm long, at the tips of which the flower-heads are borne. (Botanically, this is called a flattened corymbose panicle). The flowers appear in late November and December. The leaves are a bluey-green in colour.

Actually, the plant varies greatly in shape and form according to its locality, the leaves especially, both in size and shape, differing considerably. Its common name is the Yellow Starwort, or Ragwort, and it is known to be poisonous to cattle.

The Zulu name is insangansanga yentaba. The roots and leaves are ground and boiled by the Zulus and the unstrained decoction is then poured into a clay pot and a little taken every morning to ensure easy child-birth.

In the Drakensberg you will often come across yellow flowers, rather like a dandelion, fairly tall, with little to distinguish them to an untrained eye. They can be confusing, but most of them do at least belong to the Compositae family. *Senecio caudatus* is one of them. This is a tall plant, about a metre high, with broad, oblong leaves and a long stem, at the tip of which the flower-head is borne. It is sometimes branched. The specific name comes from the Latin word *cauda,* meaning a tail.

Just as there are many dandelion-like flowers in the Drakensberg, so there are innumerable daisy-like flowers, and it is extremely difficult to keep track of them. To add to the difficulty, most of them are yellow in colour. As an example, we illustrate *Senecio harveianus*. This is rather like an enlarged version of the common weed, known as Chick-weed *(Stellaria media)*. It grows on the summit of the Little Berg, at an altitude of about 2 000 metres, and flowers in April and May.

1

2

3

4

1. *Senecio caudatus.*
2. *Senecio bupleuroides.*
3. *Senecio harveianus.*
4. A field of *Senecio bupleuroides.*

Osteospermum is essentially an African genus, but there are two species in the Middle East and one in St. Helena. One species is endemic on Mt. Kilimanjaro. The vast majority occur in South Africa, and Natal can lay claim to ten species. It is closely related to the genus *Dimorphotheca*. In South Africa the genus occurs mostly in the south-western Cape, and in Natal along the coast and in Zululand, but I know of five species that occur in the Drakensberg.

Most of the Osteospermums are golden or orange in colour, but *Osteospermum jucundum* is a rich pinkish-mauve. It flowers in early spring, and may be found growing on the upper slopes of the Little Berg, at about 1 800 metres. The plant is about 200 mm in height, and the daisy-like flower-heads are roughly 50 mm in diameter. The leaves are quite distinctive, long and rough to the touch, tapering near the base and rounded at the tip. It is essentially a mountain flower and must not be confused with *Osteospermum barberiae,* which it closely resembles. Fortunately this latter species does not occur in the Drakensberg, being found only at lower levels in the Eastern Cape.

Osteospermum thodei, another Drakensberg species, has yellow rays and a yellow disc. It has only been recorded in a few small areas in the Drakensberg, in the Mnweni area, at Mont-aux-Sources and at Drakensberg Gardens at about the 1 800-2 700 metre levels. It flowers from November to February.

Osteospermum caulescens is rather similar to *O. jucundum,* and grows in the same area, but here the rays are white or cream above and bluish-mauve below. The disc is whitish. It is also very similar to *O. fruticosum,* but this latter species does not occur in the Drakensberg.

The word Osteospermum comes from two Greek words, *osteon,* meaning a bone and *sperma* a seed. *Jucundum* comes from the Latin word *jucundus,* pleasing or delightful. *Caulescens* comes from the Latin word *caulis,* a stalk, and means that the stalk of this species is particularly prominent. The Zulu name for *Osteospermum jucundum* is u-Mesigcolo-nkonekazi. The Zulus use it medicinally as a cure for stomach troubles. The leaves and roots are first pounded up, soaked in water, and an infusion made from them.

Osteospermum jucundum

CIRSIUM

Compositae

On a dark night early in the XIth century, so the story goes, an army of Danes was advancing on a Scottish fort, high on a hillside. In the darkness some of the men fell, unsuspectingly, into a ditch full of a plant with prickly leaves. Their shouts of pain alerted the Scots, who seized their arms and repelled the invaders. In gratitude the flower, known to-day as the Scotch Thistle, *Cirsium vulgare,* was made the national emblem of Scotland. From Scotland it has spread far and wide. It is not indigenous to South Africa, but is a common sight to-day in the lower foothills of the Little Berg. It was first recorded in Natal in 1915, at Van Reenen.

The plant is about 0,6 m high, with spiny, prickly leaves, the inflorescence a bright reddish-purple, emerging from a globose cluster of bracts. *Cirsium vulgare* is the only species we have in Natal. *Vulgare* means, in Latin, ordinary or common.

HIRPICIUM

Compositae

Hirpicium armerioides subsp. armerioides is not a very common flower, and only grows on the summit of the main escarpment, at an altitude of 3 300 metres. Again, it is the only Natal species in the genus, which is a small one, only 12 species altogether. It is found mostly in South-West Africa and Namaqualand. It used to be called *Gazania armerioides,* but the genus has recently been changed. The plant is a low-lying one, about 100 mm high, the flower-heads consisting of white rays and a yellow disc centre. It flowers from early spring into the summer.

The specific name, *armerioides* means like Armeria, a flower which is common in the high mountain regions of Scotland.

1

2

3

1. *Hirpicium armerioides subsp. armerioides.*
2. *Cirsium vulgare.*
3. *Hirpicium armerioides subsp. armerioides.*

ASTER

Compositae

The genus *Aster,* with about 250 species, is a very large one. Most of them occur in North America, but here in South Africa we have from 70 to 80 species. At least seven occur in the Drakensberg. The majority of the flowers in this genus are azure-blue in colour.

Actually there is a good deal of confusion amongst botanists about this genus. Some include the genera *Felicia* and *Diplopappus* under *Aster. Aster filifolius,* for instance, can also be called *Diplopappus filifolius,* or *Felicia filifolia.* At the National Herbarium in Pretoria *Felicia* and *Diplopappus* are grouped under *Aster,* while they are kept distinct at the Bolus Herbarium in Cape Town.

Aster filifolius (Felicia filifolia) is a small, rather insignificant shrub, about 0,6 m high, found on the lower slopes of the Little Berg and in stream beds, but in early spring it bursts into a sheet of bloom for three or four weeks. The bush has a small, dainty daisy-like flower, with purple 'petals' and yellow centre. The blooms are so abundant that they often obscure the leaves, which are smooth and narrow. It is widespread from the Cape to the Transvaal highveld, as far north as Lydenburg. It is reported to have caused fatal poisoning in sheep.

Aster perfoliatus has flowers very similar to *Aster filifolius,* but a littler larger. The plants are smaller, only about 0,3 m high, and they usually grow at higher altitudes, on the summit of the Little Berg. The flowers appear in November and December.

Aster bakerianus occurs from the Eastern Cape and on into Natal (where it is widespread) and into Swaziland and Tropical Africa. The plants are about 0,4 m high, and the leaves and stems are covered with fine hairs. The flowers are a little more blue than those of *Aster filifolius.*

Aster grauii is very similar to *Aster bakerianus.* Dr O. M. Hilliard, in her authoritative "Compositae in Natal" gives this as a syn. for *A. bakerianus,* the only apparent difference being the leaves, which are shorter and broader than those of *Aster bakerianus,* which are long and narrow. It grows on the lower slopes of the Little Berg, at about the 1 400 metre level, and the flowers appear in August. *Aster bakerianus* flowers a little later, in September and October.

RELHANIA

Compositae

Relhania tuberosa is a plant with bright yellow blooms growing on a long spike, about 0,3 m long. The flower-heads are closely compacted, and the leaves small and ericoid. It grows near the summit of the Little Berg, at about 1 800 metres, and the flowers appear during the month of June.

Relhania is a small genus, with only 18 South African species, three of which occur in Natal, the one illustrated opposite and *R. acerosa* and *R. pungens.*

---

1. *Relhania tuberosa.*
2. *Aster filifolius (Felicia filifolia)*
3. *Aster perfoliatus.*
4. *Aster grauii*
5. *Aster bakerianus.*

1

3

4

2

5

ACKNOWLEDGMENTS

One of the pleasant things about writing a book such as this is the new friendships one makes with other wild flower enthusiasts. Lovers of wild flowers are like gardeners – they are all eager to share their treasures with others. I am deeply grateful to all those many friends, old and new, who have shared their knowledge with me and who have kept me informed as to where particular flowers were to be found, and when they were flowering. I could not have written this book without them. I thank them for their help and shall always remember the pleasant companionship of those of them who accompanied me on my many trips into the mountains. Amongst them are Paul Miles and Eric Jensen of the Giant's Castle Game Reserve, and Bob Moore of Harrismith. Alfred and Jean van der Zeyde, of the Drakensberg Botanic Gardens at Harrismith, shared both their knowledge and their home with me on many occasions, as also did Bill and Leila Barnes of Giant's Castle. I think also of Cynthia Giddy, of Alex and Nancy Willcox, of Edwin and Gwen Donaldson, of Bill and Denise Carter, and of course of Godfrey and Norah Symons, the warmth of whose friendship and the lovely hospitality of whose home I shall never forget. To all these, and to many others, I offer my sincere and grateful thanks.

Bill Trauseld, whose own book on the Wild Flowers of the Drakensberg has been my constant inspiration, gave me immense help, and I also wish to thank Elsie Esterhuizen, of the Bolus Herbarium, Cape Town, for solving so many of my problems.

I am particularly grateful to Dr K. D. Gordon-Gray of the University of Natal and to the Director of the Botanical Research Institute, Pretoria, and his Staff, for the identification of my specimens. They went to endless trouble over this important aspect of my work. I must also thank the Natal Parks Board for supplying me with the necessary permit to collect specimens for identification.

Dr O. M. Hilliard, of the University of Natal, could not have been kinder in responding to my many requests for help, and she has earned my sincere gratitude for all the time she gave to me.

On several occasions in this book I have taken the liberty of quoting directly from two delightful publications on wild flowers, "Cape Floral Kingdom" by Conrad Lighton and "Wild Flowers of Table Mountain" by Professor W. P. U. Jackson. Both authors have allowed me to quote from them, and I thank them, together with their respective publishers, Juta and Co., Ltd. and Howard B. Timmins, most sincerely.

The debt I owe my son, Malcolm, is immense. Not only has he accompanied me on most of my trips into the mountains (carrying, recently, my own pack as well as his own!) but his advice on photographic matters and his constant encouragement have been an inspiration to me. 24 of the photographs in this book are from his camera. In addition to these 24 photographs, I am also indebted to the following good friends who, when I failed to obtain the pictures I wanted, came to the rescue with the offer of pictures of their own:

Muriel Zonneveld for pictures 1 and 3 on page 21
Bob Moore for the first picture on page 187
Godfrey Symons for the third picture on page 97 and No. 3 on page 197.

The beautiful line drawings, which I am sure have so delighted you, are the work of Muriel Zonneveld, a dedicated artist, keen conservationist and great lover of the mountains. I cannot thank her enough for all the long hours of work she has put into her task.

I would also like to thank my publisher, Mr Howard Timmins. During the publication of my previous book, "Barrier of Spears", I came to know him as a great and true gentleman. I thank him for all the understanding and patience he has once again displayed in this mutual venture of ours, and for all his expert advice and help.

I have already thanked Professor A. W. Bayer for so meticulously checking my MS, and Mr S. Bourquin for his advice and help over the correct spelling of Zulu names. I do so once again.

I owe a special word of thanks to Dr the Hon. P. G. J. Koornhof for his very gracious tribute to the book, on page 8.

In the final stages of the launching of the book Mr Henry Hyde of "The Estcourt Gazette" and Mr John Tungay, founder of the Drakensberg Boys Choir, played a notable and vital part. To both these gentlemen I offer my very sincere gratitude.

The typing of the MS was undertaken at short notice by Penuel Dugmore and I am deeply grateful to her, for her willingness to help and for the speed with which she accomplished her task. I can only say a simple, but very big, 'thank you' – and this comes with deep sincerity.

And finally to Edith, my wife, for all her love and understanding patience, not only while I was away in the mountains flower-hunting but during the long hours when I had to shut myself away in my study with my typewriter and my slides, another very deep, and loving, thank you. This book is dedicated to her.

THE PHOTOGRAPHING OF WILD FLOWERS

It is not for me to try and teach you how to photograph wild flowers. I have been trying to do this for over twenty years, and I am still learning! Experience is your best teacher. But perhaps a few tips which I have found useful may be of help.

The choice of a camera is yours. Basically you have an initial choice between a 35 mm camera and one of the bigger format cameras, say a 2¼" × 2¼" (6 cm × 6 cm), or even larger. In making your choice remember that the larger format camera will give you, ultimately, sharper pictures if they are to be enlarged for publication, but against that are two factors: your expenses go up, and you have an added weight to carry on your expeditions into the country, no small point when you also have to take into account tripod and other accessories such as supplementary lenses etc.

For close-up work (and it is essential that you should be able to do this) I use an SMC Pentax Macro 100 mm f/4 lens, which enables me to focus from 45 cm (1,48 feet) to infinity. For extra close-up work (photographing very small flowers) I add to this lens a set of close-up rings. I find that this set-up gives me all I require. Equally effective, of course, is an extension bellows.

The only other vital pieces of equipment that you will need are an exposure meter (if your camera is not equipped with through-the-lens metering) and a tripod. In spite of the extra weight involved, I always, wherever possible, use a tripod. A tripod eliminates camera shake at the slower speeds that stopping down entails, and it enables you so much more easily to check on your composition.

The choice of film also is yours, but it is most advisable to use a really fast film, ASA 50 as a minimum. This is because wind is your greatest enemy, and to beat it, fast speeds are essential. You will find that certain films render certain colours rather better than others. Kodachrome, for instance, gives a clearer white, I think, than Agfa, otherwise an excellent film. I would suggest that you make a series of test exposures using all the more reputable films on the market, and then select accordingly. But having chosen your film, I think it is wise to keep to it. In this way you can learn, from experience, just what it can do and what it can't.

Lighting is always a problem, for it varies so greatly. Wherever possible, I use direct sunlight, but never with the sun behind me. This latter results, I find, in flat lighting. I always try to get the sun on either my left or my right side. I find that this emphasises the shape of the flower, and helps to bring out the texture of the petals. Better still, I try to get the sun dead in front, shielding, of course, the lens. This back-lighting against a dark background (a rock or shrub in shadow) will often give you your best pictures. The light shines through the petals, picking out their texture, they literally glow with colour, and the whole flower stands out in striking contrast. The pictures on page 51, No. 1 and page 149, No. 1, were taken in this way. With side lighting dark shadows are your only problem, but this can easily be overcome by using a reflector or fill-in flash. I find a piece of crumpled aluminium foil on a stiff card most useful in this respect. Even an old newspaper will do.

I think it wise to remember one thing, though: the midday sun can be a little harsh. I have found it is better to photograph in the morning or the afternoon, but not too early or too late. The light very early in the morning is much yellower than you think, and in the late afternoon it often has a reddish cast. Both of these conditions can be fatal if you are photographing a white flower.

I find direct sunlight the best, but pale, filtered sunlight, especially when photographing a white flower, can give most pleasing results. I find that flowers photographed in shade or on a dull day lack sparkle. Here I resort to flash, either as a fill-in or 100% flash, stopping down to eliminate extraneous light. If you use flash, you should hold the flash-gun to one side, with a reflector on the other side.

At home, in the studio, I have used a simple device for flash which has given me excellent results. If you wish to try it, cut out a piece of masonite, about 50 cm × 60 cm, and cut a hole, the size of your camera lens, in the centre of the board and about 15 cm from the lower edge. Paint one side a glossy white, and mount the board vertically on your table top, with the circular hole at the bottom. Place your camera so that the lens is pointing through the hole; your camera can stand on a bracket mounted on the back of the masonite board, and it is, of course, facing directly towards the flower you are photographing. The flower stands on a sheet of white paper in front of the camera, and behind the flower, 80 cm or so above the table top, mount your flash-gun, pointing it downwards towards the flower. Your camera tripod will make an easy mount for the flash-gun. In this way you achieve back lighting, plus all-round subdued lighting, and with an almost complete absence of dark shadow. The picture on page 181, No. 1 was taken in this way. The diagram on the next page will make clear how the device works.

As an alternative, of course, you could use a series of photo-flood lamps, with the great advantage of being able to study your lighting before clicking the shutter.

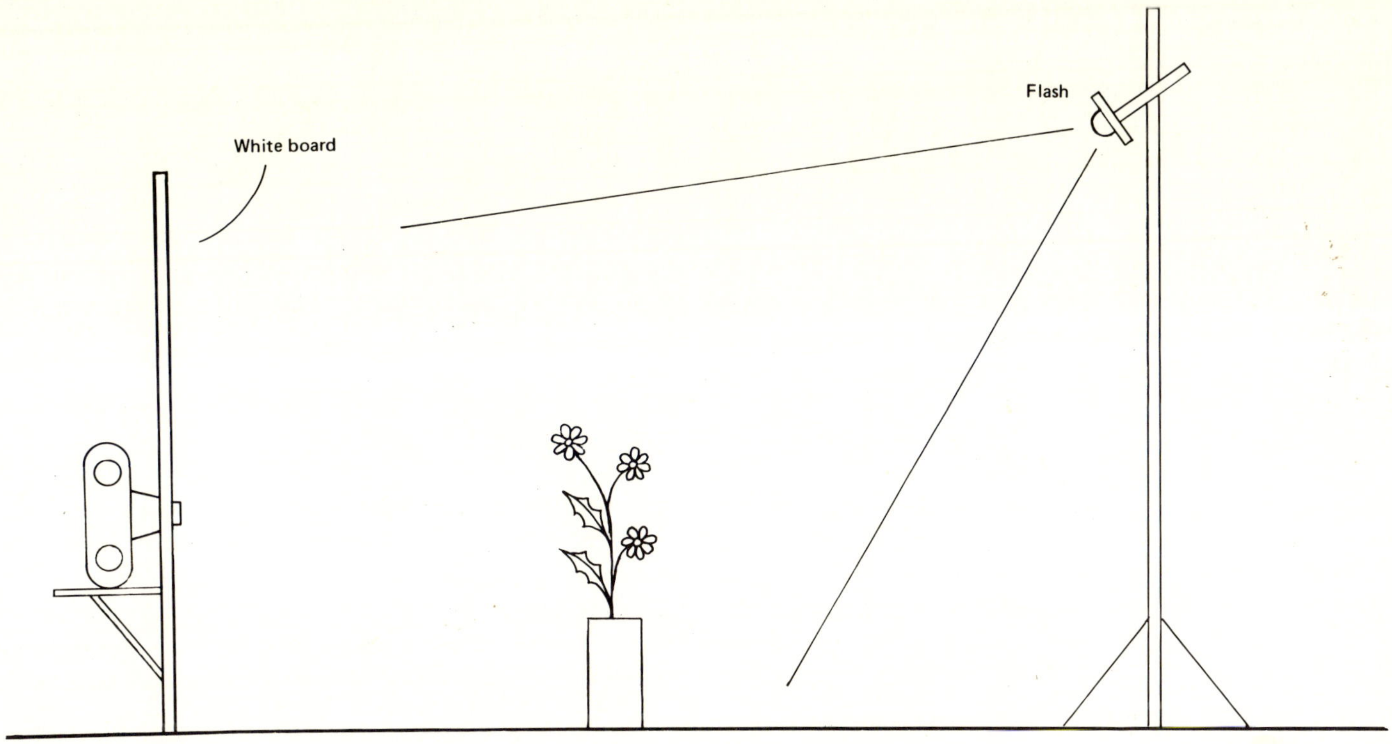

Backgrounds, I believe, are very important. Try to keep them simple, uncluttered and out of focus. Wherever possible, photograph lighter-coloured flowers against a dark background, and dark flowers against a lighter background. As mentioned above, a dark background, in shadow, thrown out of focus, invariably makes for a very pleasing effect. (Actually, as you will probably have already noticed, I prefer dark backgrounds on the whole to lighter ones). If you cannot obtain a natural background, a very useful gimmick is a series of masonite boards painted in different colours, which can be used as backgrounds. But in this case, be sure they are out of focus, or at least very smoothly painted. The lower lefthand picture on page 169, is an example of what not to do. (This was one of my earlier pictures, before I had learnt my lesson!) The sky, I find, makes a good background (see the third picture on page 199), and also water.

Wherever possible, I suggest you use the smallest stop you can, consistent with blurring the background. This will give you greater depth of focus. Of course, one should remember that if you stop down you will need a slower time, and here you run into trouble if the wind is blowing, as it always does if you want to photograph a flower in the open!

Flowers should, I think, always be photographed *in situ,* without picking them. They look far more natural if they are photographed in this way. Sometimes, however, it is necessary to pick them, as, for instance, if the background is unsuitable, or if the flower is in shade. (If you do pick it, make sure you have a permit to do this). The problem then arises as to how to mount the flower for the actual photograph. Tying it on to a stick is always clumsy, and the flower is never just as you want it. Here an excellent tip is to carry with you a lump of child's plasticine. Attach a small piece of this to the stalk of your flower, and you can then mount it anywhere, to a stick driven into the ground or against the trunk of a tree, and it is so easy to manipulate the flower into just the position you want. Another method is to use the stem of an old clip-on reading lamp with the head removed. In its place solder a bull-dog type clip, into which the flower is fastened. This, again, gives complete freedom of movement, as the stem can be moved in any desired way.

In photographing the flower *in situ* wind is often a problem. This can sometimes be overcome, in the case of a tall flower, by driving a stick into the ground near the flower, and either tying the flower to the stick, or, again, using your lump of plasticine. Or use flash. Flash goes off at about 1/2 000 sec., sufficient to freeze even the most violent movement. Another useful way to beat the wind is to pick the flower, take it indoors, and photograph it in front of a closed window through which the sun is shining. A reflector card on the other side is essential, however, if you do this. The picture on page 77 (No. 1) was taken in this way, but I forgot the reflector card, with consequent dark shadows on the lefthand side.

Photographing blue flowers can be a real problem.

They rarely come out with the right shade of blue. I have even, on occasion, found a blue flower coming out a bright pinkish-mauve! Gentians come out pink, and Blue-bells and Aristeas with a mauve tint (Strangely enough, the older, original colour films did not have this defect. The picture of the Aristea, on page 81, was taken many years ago, with the old, original Kodachrome film, and the colour is true. Other pictures of Aristea, taken recently, I have had to discard for the purposes of this book, for the colour was all wrong). It is thought that some blue flowers radiate infra-red rays to keep themselves cool. These rays, picked up by the camera, give a reddish cast to the final picture. I have tried two ways of solving the problem. Either use a blue filter, making sure first of all that the background is sufficiently dark, or give your final processed transparency a wash of blue colour from a Johnson's Colouring Outfit, again making sure that the background is a dark one. Note the first picture on page 187, a good example of this difficulty. The colour of this flower is actually blue.

If you want to preserve the fresh, dewy look of a mountain flower, try spraying it with water. If you haven't a spray-gun handy, fill your mouth with water, take a deep breath, and spurt the water over the flower in a fine spray! The top picture on page 175 was taken in this way.

You will, of course, want to study your composition, and this is not easy. In fact, it can be quite a problem at times. The flowers never seem to quite arrange themselves as you want them. One good tip is to try and limit the flowers in your picture to odd numbers, (ones, threes or fives, for instance) rather than even numbers. If you have to photograph two flowers together, I suggest that one should be placed at a slightly lower level than the other. Compare, for instance, the picture on page 127 (No. 1) with the one on page 1 (No. 5) where the latter picture is not as good as the former one. This also applies to a row of flowers. I think we should always try and avoid a straight horizontal line of flowers, evenly spaced. Try and break the line, placing some flowers at a higher or lower level. If you are photographing one flower alone, with its stalk, I think it is wise to try and avoid photographing the flower standing up at right-angles to the side, in the centre of the picture. Try the effect of a diagonal entrant for the flower, with the stem emerging from one of the lower corners, instead of from the middle of the lower margin. I leave you, the reader, to decide which is the better of the two pictures, No. 4 on page 95, or No. 2 on page 115. Summarising then:

1. Move in close.
2. Use side-lighting or back-lighting.
3. Watch your backgrounds. Keep them simple and uncluttered.
4. Use a reflector or fill-in flash to lighten the shadow areas.
5. Sunlight is always best.
6. In the absence of sunlight, use flash.

The photography of wild flowers can be a very exciting hobby. These are but a few tips which I hope you will find useful. Good shooting to you!

GLOSSARY

Anther	The part of the stamen which contains the pollen.
Awn	A fine bristle, usually at the end of an organ.
Axis	The angle between the leaf and the branch.
Bisexual	A flower which has both stamens and pistil, i.e. with both male and female parts.
Bract	A modified leaf found at the base of a flower, or flower stalk intermediate between the calyx and the normal leaves.
Calyx	The outer envelope of the flower, consisting of the sepals.
Campanulate	Bell-shaped.
Capitulum	See below, Types of Inflorescence.
Conical	Cone-shaped, like a carrot.
Corm	Rather like a bulb, but with a solid centre, and more compressed. New corms are often found on top of the old ones.
Corolla	The inner envelope of the flower, consisting of petals.
Corymb	See below, Types of Inflorescence.
Crenate	Crinkly, indented, notched with rounded teeth.
Cyme	See below, Types of Inflorescence.
Dioecious	With unisexual flowers, the male flowers quite separate from the female flowers, and growing on separate plants.
Disc florets	The central florets in the capitulum of a Compositae, as compared with the outer ray florets.
Endemic	Confined to a particular region or country, and found nowhere else.
Epiphyte	A plant which grows on other plants without obtaining nourishment from them.
Ericoid	With leaves like those of an Erica – small, narrow and folded.
Exotic	Introduced from abroad. The opposite to indigenous.
Filament	The thread-like stalk or stem of an anther.
Glabrous	Devoid of hairs, smooth.
Globose	Shaped like a globe. More or less spherical.
Herbaceous	A plant with a soft stem (not woody) coming up annually from a persistent root.
Imbricate	Overlapping, like the tiles on a roof.
Inflorescence	The arrangement of the flower parts. Less correctly used for the flower itself.
Indigenous	Original to the country, not introduced.
Involucre	A number of bracts surrounding the base of a flower-head.
Keel	A ridge, like the keel of a boat.
Labellum	Lip, the lowest and largest of the three inner petals of an orchid flower.
Lamina	The blade of a leaf.
Lanceolate	Narrow, tapering at each end.
Latex	Milky juice.
Maculate	Blotched or spotted.
Monoecious	When the male and female flowers are separate, but borne on the same plant. (See Dioecious).
Obconical	Conical, but attached at the narrower end. (Ob- is a prefix indicating the reverse of the following word).
Orbicular	Flat, but round.
Ovary	That part of the pistil (q.v.) which contains the ovules (the immature fruit).
Ovule	The immature fruit, or seed, in the ovary, before fertilisation.
Panicle	An inflorescence consisting of branches bearing several flowers, a cluster of flowers, a branched raceme.
Pappus	The ring of hairs at the top of a seed, as in *Senecio* flowers. Thistle-down.
Parasite	A plant which derives some or all of its nourishment form another plant.

Pedicel — The stalk of each individual flower.

Peduncle — Stalk of an inflorescence, or a cluster of flowers.

Perianth — The floral envelope, i.e. both the calyx and the corolla.

Petaloid — Resembling a petal.

Pistil — The female part of the flower, consisting of ovary, style and stigma.

Pubescent — Covered with short, soft hairs.

Raceme — See below, Types of Inflorescence.

Ray florets — The outer florets of a Compositae, as compared with the disc florets.

Rhizome — An underground creeping stem, distinguished from true roots by the presence of buds, leaves or scales, and sending up stems.

Scape — A leafless flower stem arising from the ground.

Sepal — Each segment of the calyx.

Spadix — A spike with a fleshy, succulent axis.

Spathe — A large bract enclosing a flower cluster, as in the Arum Lily.

Spike — An inflorescence consisting of a single main stem on which the stemless (sessile) flowers are arranged.

Stamen — The male part of a flower, consisting of anther and filament.

Stigma — The tip of the pistil, where the pollen is deposited.

Style — The narrow upper part of the ovary, supporting the stigma.

Tessellate — A chequer-pattern.

Tomentose — Covered with short, soft hairs.

Tuber — A thickened underground branch or stem, producing buds.

Tunicated — Covered with a coat, or a layer of scales, as in an onion.

Umbel — See below, Types of Inflorescence.

Whorl — The arrangements of similar parts of a plant or flower in a circle around the stem or axis.

MONOCOTYLEDONS AND DICOTYLEDONS

Flowering plants are classified as either Monocotyledons or Dicotyledons.

Monocotyledons have one leaf arising from the germinating seed, storage organs like bulbs or corms, and the leaves are long, sword-like and parallel-veined. The flower parts are usually in threes or multiples of three.

In *Dicotyledons* two leaves grow from the germinating seed, the veins of the leaves (which are usually more rounded) form a network, and the flower parts are in fives and fours.

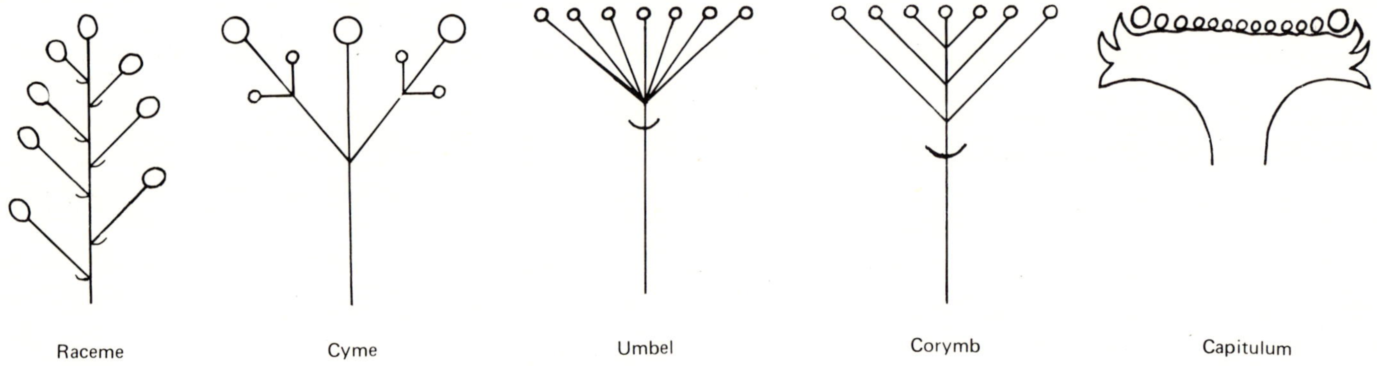

TYPES OF INFLORESCENCE

INDEX OF FAMILIES

INDEX OF BOTANICAL NAMES

(Names in italics indicate species relegated to synonymy)

INDEX OF COMMON NAMES

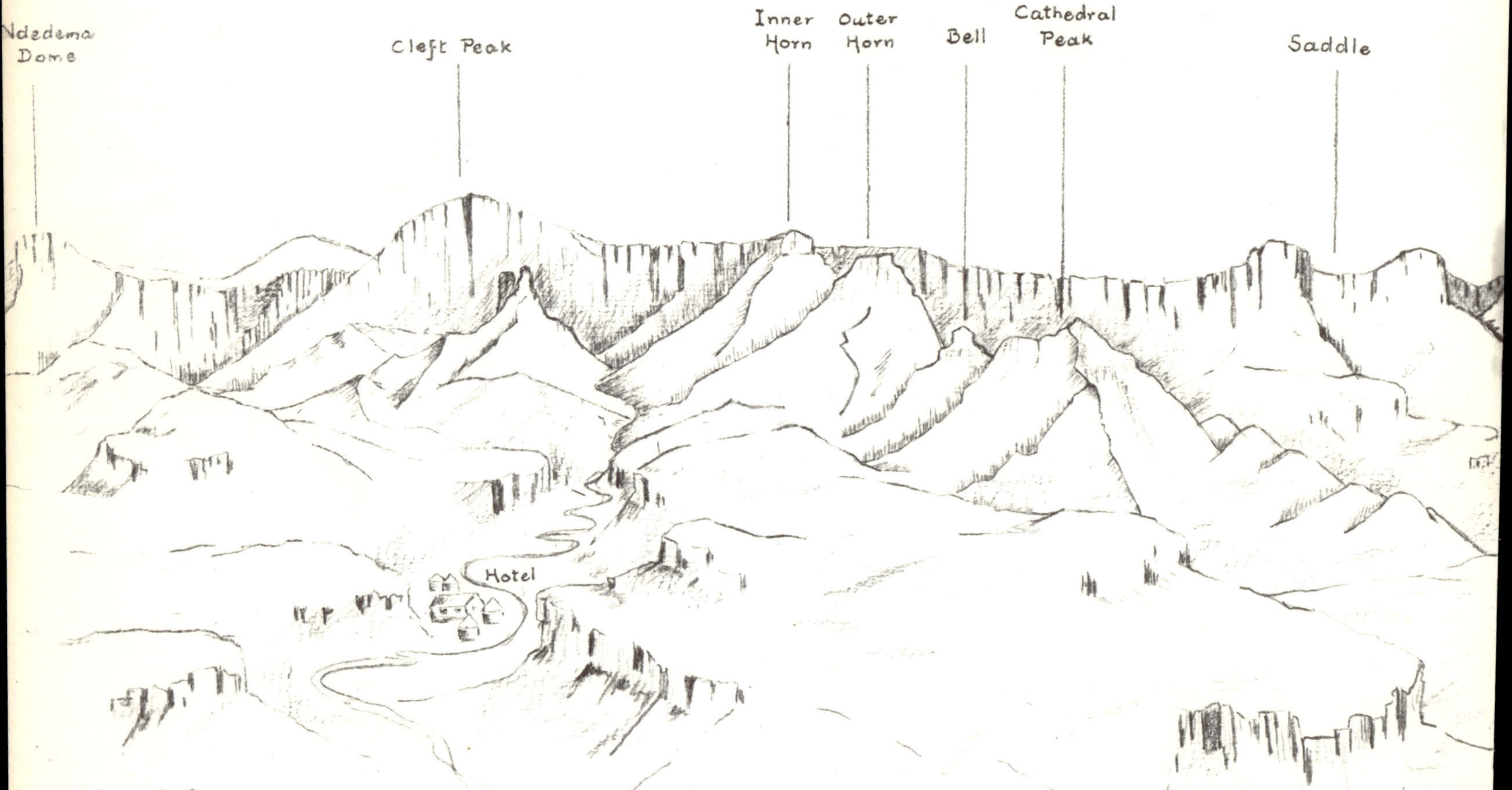

Continued from front cover →